POLITICS AND THE LIFE SCIENCES: THE STATE OF THE DISCIPLINE

RESEARCH IN BIOPOLITICS

Series Editors: Steven A. Peterson and Albert Somit

Recent Volumes:

RESEARCH IN BIOPOLITICS VOLUME 12

POLITICS AND THE LIFE SCIENCES: THE STATE OF THE DISCIPLINE

BY

ROBERT H. BLANK
University of Canterbury, Christchurch, New Zealand

SAMUEL M. HINES, JR.
The Citadel, Charleston, SC, USA

ODELIA FUNKE
US Environmental Protection Agency, Washington, DC, USA

JOSEPH LOSCO
Ball State University, Muncie, IN, USA

PATRICK STEWART
University of Arkansas, Fayetteville, AR, USA

United Kingdom – North America – Japan
India – Malaysia – China

Emerald Group Publishing Limited
Howard House, Wagon Lane, Bingley BD16 1WA, UK

First edition 2014

British Library Cataloguing in Publication Data
A catalogue record for this book is available from the British Library

ISBN: 978-1-78441-108-4
ISSN: 2042-9940 (Series)

CONTENTS

EDITORIAL ADVISORY BOARD

PREFACE

In this book, we examine the emergence of biopolitics as an academic perspective within the discipline of political science over the last four decades and review the rich literature of biopolitics as represented by the leading proponents of the perspective. Much of the early work in biopolitics was an attempt to convince political scientists on the importance of a broad Darwinian approach to politics – a needed shift in paradigm away from the predominant standard social science model. Largely, the message fell on deaf ears for many decades, however, in the last decade its marginalization has begun to weaken.

In part the changing status of biopolitics is reflective of a more general pattern in academia, where biological explanations of social phenomena progressively have been viewed as acceptable and even necessary. It has also been facilitated by methodological developments in the life sciences, especially behavioral genetics and cognitive neuroscience, which permit the incorporation of biological variables into studies of social behavior. In addition, this new more receptive atmosphere has been facilitated by a theoretical shift within the biopolitical literature itself where earlier attempts at grand theorizing are being superseded by efforts to use lower-level biological theories to derive and empirically test hypotheses on political behavior. We shall examine this changing scholarly environment and review the rapidly expanding work of the "second generation" biopolitics political scientists as well as relevant work from life scientists and others.

The authors would like to thank Steven A. Peterson and Albert Somit for their strong support of this project and Thomas Dark at Emerald Press for agreeing to publish an authored volume in a series that until now consisted of edited volumes. Although the book represents a joint effort of the five authors, on the basis of their expertise each took the lead on at least one chapter. These responsibilities included: Joseph Losco, biopolitical theory; Patrick Stewart, biopolitical methodology; Odelia Funke, biopolicy; Samuel M. Hines, Jr. comparative/international and environmental/ecological issues; and Robert H. Blank, political behavior. All five authors represent early first generation biopolitics scholars who look forward with anticipation to the work of the second and future generation scholars that builds upon the contributions of our many colleagues in biopolitics to whom we owe a debt of gratitude and dedicate this volume.

Robert H. Blank
Samuel M. Hines, Jr.
Odelia Funke
Joseph Losco
Patrick Stewart

CHAPTER 1

BIOLOGY AND POLITICS: AN INTRODUCTION

Robert H. Blank

Until recently, the notion that political attitudes and behaviors are in part a result of differences in genetic structure, neurological function, hormones, and health in general was not popular in political science. However, it is no longer possible to deny this as mounting evidence demonstrates the importance of the combination of both our genetic and physiological disposition in conjunction with the environments within which we live. In more integrative approaches, the social environment and individual disposition are mutually incorporated into understanding a trait. This integrative view has now begun to move into the social sciences (Hatemi, 2011). Thus, evolutionary biology is "increasingly serving as a wellspring of hypotheses in the social sciences, especially economics and psychology, but also in political science" (Smith, Larimer, Littvay, & Hibbing, 2007). However, despite the discernible implications of this research for our understanding of human behavior, we are still far from realizing its fundamental importance to political science.

Over the past four decades, there has been a mounting interest among social scientists in the theories, research, and findings of their counterparts in the life sciences. This increased level of interest manifests itself among social scientists interested in explaining human social behavior and the functions and operations of social institutions as well as among those

Politics and the Life Sciences: The State of the Discipline
Research in Biopolitics, Volume 12, 1–34

ISSN: 2042-9940/doi:10.1108/S2042-994020140000012001

interested in public policy issues including the environment, health, and economy. The prefix *bio* is being placed on politics, economics, and sociology with ever-greater frequency in scholarly publications. Moreover, the recent progress of evolutionary psychology and behavioral genetics manifests the representation of the social sciences in an interdisciplinary biosocial science upsurge. Anthropology (especially physical anthropology) and geography have long been sensitive to the life sciences. Although psychologists, long dominated by Skinnerian behaviorism, have been more attuned to biology than social scientists at least since the revival of cognitive studies in the 1950s, recently their interest in neurophysiology and psychobiology has grown appreciably.

In this book, we analyze the emergence of politics and the life sciences, also termed biology and politics or simply biopolitics, as an academic perspective within the discipline of political science; review the biopolitical literature as represented by the leading proponents of the perspective as well as others; and present a coherent view of a biological-based framework to structure political inquiry. Although the term biopolitics serves as a short-hand for this perspective, as we will see shortly, its use is not without controversy. Whatever it is called, however, this new integrative approach offers political science a new *naturalism* that will propel it beyond the sterile separation of facts and values and challenge it to move decisively beyond the limitations of modernism with its emphasis on disembodied rationalism. Therefore, we seek to demonstrate the value of a life sciences perspective for the study of all aspects of politics, political behavior, and public policy.

POLITICS AND THE LIFE SCIENCES

This book is a reflection of the significant body of scholarly work that has been produced by a variety of political scientists since the late 1960s: a body of work that collectively offers a paradigm shift in political science that advocates the naturalistic study of politics. A biological approach to political science, of course, is hardly new and at its base reflects the tradition of inquiry in political theory that is traceable back to Aristotle, most particularly his conceptualization of man as *zoon politikon* – the political animal (Arnhart, 1988, 1992; Wahlke, 1979). As Roger Masters has argued, the insights of the contemporary life sciences require that we reexamine the ancient philosophers' view of science and recognize that some of the premodern notions they held are actually more consistent with the most contemporary thinking than the mechanistic, dualistic modern positivistic science associated with Descartes, Copernicus, Newton, and prerelativity physics (Masters, 1993).

Despite its roots in the classics, the most important theoretical source for this paradigm shift is contemporary neo-Darwinian evolutionary theory. Moreover, ethology and the emergent disciplines of behavioral ethology, human genetics, cognitive psychology, and neurobiology have much to contribute to a fully developed naturalism in the social sciences. Finally, this new naturalism in political science is indebted to advancements in contemporary physics and to chaos theory (Masters, 1993; Schubert, 1989). Overall, then, the new naturalism reflects our evolving understanding of the nature of science, the nature of explanation, and the nature of causation (see Masters, 1993). Hopefully, the significance of these new intellectual developments will become apparent as the foundations of the biopolitical movement are presented here.

This more inclusive model stresses on the powerful connections between biology, and the life sciences more generally, and politics. To understand politics, it is necessary to apply the knowledge generated through the study of the natural world. It is also essential that we recognize humans are a part of – not apart from – that natural world. Therefore, the study of politics should encompass the full range of political behavior among those species that are capable of exhibiting analogous behavior. It must also include a wide range of policy issues and topics, including the environment, human health, biotechnology, neuroscience, economic development and sustainability, population dynamics, and chemical and biological weaponry, among others. Indeed, we will argue that the growing list of significant policy issues that involve nature and human nature loom as of such importance that they should be acknowledged as at the top of humankind's political agenda. Addressing these policy issues in this broader framework is the key to our survival as a species.

The life science paradigm points toward the evolutionary history of the human species as a source of knowledge and reflects the continuity that exists from early hominids and man's closest primate relatives to the present. As noted by Harold Lasswell, one of the most interdisciplinary of 20th century social scientists, in the first issue of *Comparative Politics*, we must take a very long historical perspective if we are to understand fully the roots and origins of human political behavior and the products of that behavior, political institutions, and processes (1968). Glendon Schubert reiterated this point dramatically:

> Biological theory implies the rejection of the presumption that our political theory as a species began 2,500 years ago in Athens or (alternatively) as described in "naturalistic"

> fables (whether optimistic like that of Rousseau or pessimistic like that of Hobbes).... The roots of political behavior go back not thousands but millions of years; and political man did not spring ... from the forehead of Socrates – as our teaching of the wellsprings of political philosophy might lead innocents to infer. The implications of contemporary research in physical anthropology, archeology, paleontology, and related sciences are going to jack political philosophy off its classical assumptions – once political scientists become better educated in, and start facing up to the facts of biological life including their own life history as a species. (1976, pp. 164–165)

Despite differences in their specific interest areas, proponents of political science that is informed by the life sciences are unified by their advocacy of the application of the knowledge and methods of the life sciences to the study of politics, and their rejection of a narrow, positivistic philosophy of social science and, particularly a simplistic, reductionist science. Rather, they seek to understand the biological bases of behavior and to comprehend the theoretical basis of political theory as an enterprise that is both empirical and normative in its scope.

Advocates of this approach are also highly committed to the interdisciplinary study of politics. The greatest challenge posed by this paradigm shift is the requirement that practitioners of political science must include the theory, methods, and findings of the life sciences in their work. Arguably, supporters assert that some of the most important insights into political behavior are to be found outside of the mainstream of political science. John Wahlke, in his Presidential Address to the American Political Science Association (APSA), echoed this point of view when he challenged the profession to move beyond a "pre-behavioral" study of politics to a "more genuinely behavioral political science" (1979, p. 30). More recently, Nobel Laureate Elinor Ostrom in her 1998 APSA Presidential Address invoked evolutionary theory 10 times while addressing collective action and rational choice. Overall, however, despite increasing evidence of the importance of biology to human behavior, social scientists in general, and political scientists in particular, have not been inclined to embrace this new naturalism. As Masters observes: "Most social scientists, when encountering the evolutionary perspective on human affairs, have ignored it or tried to suppress it, using the anonymity of peer review to oppose the publication of scholarly work they neither understand nor respect" (1993, p. 143).

It should be noted, however, that political science has not always been hesitant to break new ground, particularly in its relationship to biology. In fact, there has been a progressive narrowing of the field since its formative period in the earlier decades of the 20th century. A reading of the presidential addresses and articles of the "Founders" of the APSA demonstrates a robust interest in the interdisciplinary goals of the discipline, especially as it related to the life sciences. The following quotations illustrate this early concern. William Bennett Munro, in his 1927 Presidential Address to the APSA, argued that:

> Our immediate goal, therefore, should be to release political science from the old metaphysical and juristic concepts upon which it has traditionally been based ... *It is to the natural sciences that we may most profitably turn, in this hour of transition*, for suggestions as to the reconstruction of our postulates and methods [Emphasis added]. (Munro, 1928, p. 7)

Perhaps the clearest statement of the special reciprocal relationship between the natural sciences and the social sciences comes from Charles Merriam in his 1925 APSA Presidential Address:

> Still more serious for the student of politics is the integration of social science with the results of what is called natural science – the reunion of the natural and the "non-natural" sciences for more and more it appears that the last word in human behavior is to be scientific; more and more clearly it becomes evident that the social and political implications of natural science are of fundamental importance. It even seems at times that this is more evident to the natural scientists than to the social scientists, who at times concede the impossibility of more scientific social control of human conduct. (Merriam, 1926, p. 15)

Little can be added today to strengthen what Merriam said almost 90 years ago. Within the context of the emerging revolution, especially in genetics and neuroscience, Merriam's observations are even more relevant today. Moreover, the opportunities that Merriam foresaw for political science have increased dramatically as have the long-term deleterious consequences for a failure to respond.

In 2010, a workshop funded by the National Science Foundation's (NSF's) Political Science Program concluded that there are many potential benefits of a biologically enriched social science. Such knowledge can give us greater insight into the human condition and has the potential to inform public policy.

> While many people offer opinions about the causes of human social behavior, it is critical that scientific inquiries into such matters be supported ... societies that have the ability to confront important social hypotheses with credible evidence and rigorous evaluative procedures also enhance their capacity to develop and maintain increasingly complex social institutions ... The effective and efficient functioning of numerous aspects of modern social infrastructure depends on the insights that science can offer. (Lupia, 2011a, p. 86)

HISTORICAL DEVELOPMENT OF BIOPOLITICS

Although the term biopolitics was used by Rudolf Kjellen, a Swedish scholar who saw the state as a life form that preceded individuals, as early as the 1920s (Lemke, 2010) and by Canadian novelist Morley Roberts (1938), the first political scientist to use the term in print was Lynton Caldwell who discussed the impact of biological information on the understanding of political phenomena and public policy (Caldwell, 1964). Also during the 1960s, although not using the term, James Davies (1963) was arguing that the improvements in and insights into psychobiology, neurophysiology, and endocrinology could aid facilitate an understanding of human nature and political behavior while Albert Somit (1968) was calling on political scientists to reorient their approach to political behavior by studying ethology and psychopharmacology instead of looking only at learning and environmental factors.

The term biopolitics was again used by Thomas Thorson in his book *Biopolitics* where he argued that political scientists' understanding of science, based largely on a mechanistic model drawn from 19th century physics, was inadequate because it did not consider nature and human development, time, and change. He defined biopolitics as "politics understood by man as evolution becomes conscious of itself" (1970, p. 210). In 1975, at a conference in Paris sponsored by the International Political Science Association (IPSA), a group of political and social scientists gathered to present papers on biopolitics that were later published in a volume edited by Somit entitled *Biology and Politics* (1976). Furthermore, a special issue of *The International Political Science Review* was devoted to articles on biopolitics. Moreover, a Biology and Politics section of *Social Science Information* was long edited by Roger Masters, another key player of the biopolitics movement in the United States. Over the decades, an important vehicle for publishing biopolitics-based research has been the series of volumes, *Research in Biopolitics*, edited by Somit and Peterson.

By the 1970s, four major strands of biopolitical publications were apparent: essays making the case for a more biologically oriented political science; ethological and evolutionary analyses to develop broad-scale understandings of political behaviors; research on the physiological and pharmacological antecedents of political behavior and studies of public policies with biological or environmental connections (Somit & Peterson, 1998). With the formation of the Association for Politics and the Life Sciences (APLS) in 1980, the term biopolitics was employed to describe the interdisciplinary approach of scholars in political science who held that data and theories from the life sciences, as well as biological research techniques, could lead to a more comprehensive knowledge of political behavior. Instead of viewing human behavior as merely rational, psychological, or cultural, they urged political scientists to examine the role of biological factors in human behavior and politics.

Despite this early work in what they label "scientific biopolitics," Liesen and Walsh (2012, 2013) suggest that the term now has incompatible meanings in political science. When the term was first used among U.S. political scientists in the 1970s, it referred to what they had hoped would be an emerging subfield that incorporated the theories and data of the life sciences into the study of politics, political behavior, and public policy. Simultaneously, however, European scholars were using the term to refer to problems raised by biotechnology and to the analysis of how humans were being objectified by the life sciences. Moreover, in the late 1970s, French postmodern philosopher Michael Foucault interchangeably used the terms biopolitics and biopower to designate strategies of power that are pervasive in modern society. Consequently, according to Liesen and Walsh, there are now two competing and contradictory usages of biopolitics: scientific biopolitics that examines the theories and data of the life sciences to help understand politics, public policy, and political behavior and postmodernist biopolitics that condemns the historical manifestations of power on individuals and society.

Although Rebecca Hannagan (2013) agrees that the distinction between the two camps is accurate, she contends that it is not necessary to try and reclaim ownership from the postmodern followers or utilize alternative labels such as neuropolitics because the current scientific scholars with biological approaches to politics do not identify with the term biopolitics anyway, opting instead for evolutionary, neo-Darwinian, or ethological to describe their research. Although the term biopolitics may have been strategic for bringing a degree of coherence to the emergent and fragmented body of work of the early scientific biopolitical scholars, it is no longer necessary. Stewart (2013) adds that while ceding the term biopolitics to postmodernists might be discouraging to some, biopolitics has fallen out of use while evolutionary theory is becoming more relevant. Moreover, the appropriation of the term biopolitics by those who use it as a postmodern construct is not necessarily a bad thing since the term was mostly used colloquially by its practitioners to refer to a range of life science issues embracing biobehavior, biopolicy, and bioethics with APLS functioning largely as an umbrella under which all these life science issues could be studied. Stewart's content analysis demonstrates a concurrent increase in the use of evolutionary theory not only in political science but also in public policy and psychology, suggesting that biopolitics is thriving, albeit under a different label. Similarly, Bucy (2012) suggests that while an updated interpretation of the term must now incorporate critical or postmodern approaches, it's time to move on. Despite the disparate meanings of biopolitics in political science, research activity is thriving.

The expansion of creative research over the last decade suggests that scientific biopolitics is maturing into a more specific research area that utilizes data and theories of the natural sciences under a variety of monikers including neuropolitics, cognitive neuroscience, social neuroscience, political psychology, evolutionary politics, genopolitics, and Darwinian politics. Alford and Hibbing (2008) contend that, to its detriment, the early biopolitics movement followed the lead of evolutionary psychology and sociobiology and focused on the role of biology in establishing broad human universals such as hierarchy, war, and leadership behavior. In contrast, they maintain that a "new empirical biopolitics" is replacing the older form that was largely theoretical, descriptive, and speculative (Liesen & Walsh, 2013). This new empirical research on the biology of social behavior can be roughly divided into work that is focused on the brain and associated neurotransmitters and that centered on genes and gene markers. According to Alford and Hibbing (2008), despite the presence of a biopolitics subgroup for over 30 years, there has been little empirical work of either kind in political science.

Gary Johnson (2011), however, takes umbrage with the suggestion that biopolitics has in the past not extended beyond "think pieces" and "calls to arms" (Alford, Funk, & Hibbing, 2005, p. 199). He notes that given the extensive data-based work of Jim Schubert, Roger Masters, John Wahlke, Milton Lodge, Douglas Madsen, Meredith Watts, Tom Wiegele, George Marcus, and others, it is inaccurate to suggest that past work in biopolitics has been mere rumination. Although Alford and Hibbing (2008) later cite some of this past work, what they cite indicates that they limit empirical biopolitics only to studies that involve measurement of genetic, neurophysiological, or physiological variables. Johnson contends that empirical research using ethological methods and scientific work in primatology, for instance, should be considered empirical biopolitics although much of that work did not originally involve direct measurement of biological variables. Any empirical research that tests hypotheses drawn from empirical biopolitical theories is empirical biopolitics even when they are tested using fairly conventional political science data without direct measurement of any biological variable (Johnson, 2011).

ORGANIZATIONAL FOUNDATION OF POLITICS AND THE LIFE SCIENCES

The Center for Biopolitics was established at Northern Illinois University by Thomas Wiegele in 1977. The mission of the Center was to "establish itself as a leading focus for biologically oriented political scholarship" and it was to "play a leading a sympathetic role" "in promoting biopolitics as an area of political inquiry" (Bonnicksen, 2013). Wiegele was able to secure funding from the Scaife Foundation to bring several postdoctoral faculty members to the Center each year during the early 1980s. The Center also published a newsletter and produced regular bibliographies for several years. In 1982, the APLS was founded and it published the inaugural issue of the journal *Politics and the Life Sciences*. The journal has been published continuously since then and is currently published biannually by Allen Press in England under the sponsorship of the Association. In 1986, the Center was renamed the Program for Biosocial Research although it no longer exists as a separate administrative unit. In June 1990, *Biopolitics and Mainstream Political Science: A Master Bibliography* was published by the Program.

One of the more interesting features of the development of the early scientific biopolitics movement was the extent to which it was an international endeavor. Among the earliest conferences to sponsor panels on biopolitics was the IPSA. In 1975, this was strengthened when IPSA granted Research Committee #12 (Biology and Politics) formal status. Since then, the Committee has been dedicated to "exploring the linkage between biology and politics" through a variety of research foci that include policy issues emerging from the life sciences; the evolutionary bases of human political behavior; physiological influences on political behavior; and the development of research methods to analyze the linkage between biology and politics. Over the past four decades, the Committee, with a membership from over 40 countries, has been active in organizing panels to support this research (Somit, 2013a, 2013b).

Although not using the name biopolitics, other related academic programs have been initiated in the United States over this period. In 1968, the University of New York at Stoney Brook Political Science Department was mandated to specialize in behavioral research and experimental methods. Under this mandate, in 1977, a PhD program in political psychology was instituted. Their 2014 call for applicants for the PhD program with "research interests at the intersection of evolutionary psychology and political science" cites active research areas in cooperation, generosity, morality, alliances, property, risk preferences, agent-based modeling, and evolutionary game theory. The Department also recently initiated an interdisciplinary Center for Behavioral Political Economy (Lodge & Somit, 2013).

Other relevant programs include Vanderbilt University's Department of Political Science Research on Individuals, Politics, and Society initiative. According to Hannagan (2013), it is an example of the move toward a scientific biopolitics approach to the study of political attitudes and behavior, although wholly within the behavioralist tradition and under the subfield of political psychology. Similarly, the Department of Political Science at the UC Merced offers an interdisciplinary program incorporating cognitive science, psychology, economics, sociology, and political science to study political psychology or political economy. Subfields include political cognition and behavior and political institutions and political economy and the method sequence includes computational modeling and experimental methods in addition to more conventional methods. Similarly, The University of Nebraska offers a program in biology and politics – the Systems Biology of Social Behavior Initiative – that comprises an interdisciplinary group of researchers from psychology, political science, biological sciences, sociology, economics, as well as special education and communication disorders.

In 1985, the APLS became an organized section of the APSA. It was active in organizing biopolitics panels at the national convention for a decade, but in 1995 lost its status as an organized section when the APSA raised the minimum number requirement (for a more in-depth historical narrative, see Johnson, 2011). The changing membership of APLS, of which over half were drawn from other disciplines by the mid-1990s, meant that many members were not willing to pay full APSA membership fees to attend a few APLS panels. As a result in 1998, the Association began hosting its own annual meeting. Initially, they were held in the same city and time as the APSA to enable members to attend both conventions, but in recent years they have met separately at a different time of the year. The full programs of these meetings can be found at the APLS website (www.aplsnet.org).

Despite a relatively productive organizational history, or perhaps in part because of its success in forging an interdisciplinary association, Somit and Peterson concluded that in terms of intellectual impact on the discipline, the APLS could not be viewed as a success (1998, p. 569). In their valuable review article of biopolitics after three decades, they reluctantly came to the conclusion that with few exceptions, primarily in international relations, biopolitical research did not have an appreciable impact on political science. As measured by a "near total absence" of articles or reviews of biopolitical research or ideas in the leading journals and little mention in the major textbooks, there was scant evidence of a transition from an active interdisciplinary agenda to mainstream political science. Interestingly, Somit and Peterson noted that the picture was different if one focused instead on the wide application of biological concepts to the study of politics by those outside the discipline. Here, they found considerable activity especially by nonpolitical science trained members of the APLS.

Three decades ago, Thomas Wiegele in his seminal work *Biopolitics* was forced to conclude that at the "present state in the development of the discipline, we can characterize ourselves as having ignored an enormous amount of information about humanity and its real nature, that is its nature *as it is lived*" (1979a, p. 145). A decade later, Roger Masters, similarly concluded "the social sciences remain largely untouched by research in the biological sciences" (1989, p. xii). Even as late as 2001, Blank and Hines (2004) found that despite significant work in biopolitics over the past two decades, Master's conclusions remained accurate particularly in regard to political science. In fact, within the context of what was transpiring in other social sciences, particularly anthropology and psychology where evolutionary perspectives and biological grounding of behavior became well established, political science lagged even further behind.

This absence of a disciplinary interest in biopolitics was not surprising since political science during this period was not sensitive to the contributions of the life sciences to political life. The major contending theories for the study of politics all fail to incorporate the biological dimensions that are essential to a full understanding of political phenomena. Indeed, the major theories, particularly rational choice with its rational, self-interested economic man as the centerpiece and political socialization grounded in Skinnerian behaviorist psychology, are largely antithetical to a life sciences approach. As a result, political science in the last half of the 20th century was dominated by a "profound anti-evolutionary bias" (Philips, 1999, p. 535). In 1989, John Wahlke chaired the Political Science Major Task Force, which explored a core program of courses essential for training a political scientist. One of its findings was that "A 'new' evolutionary biology has revolutionized thinking about the formation of social aggregates by primates and hominid ancestors of modern mankind" (Wahlke, 1991, p. 52). Members cautioned that rapid advances in knowledge across a broad range of biobehavioral sciences threatened to make political science obsolete if it did not impart such knowledge to its students. The swift and hostile reaction to this recommendation demonstrated the intensity of opposition to biobehavioral initiatives. For instance, in the same issues of *PS* that carried the Task Force Report, Warren Miller charged that biology is "largely irrelevant to the center of gravity of political science," that the link between biology and the social sciences was "scarcely visible" and that studying biological processes is "quite peripheral to the interests and needs" of social scientists (1991, p. 9).

The threats of an evolutionary-based political science to the standard social science model (SSSM) are understandable since it requires the study of actual behavior, not merely attitudes and opinions, and it refuses to accept the truncated account of human nature that a rational man model relies upon for explanation. Despite continued shortcomings of the SSSM discussed later in this chapter, the case for a biobehavioral model gained little ground against the prevailing theories of politics for nearly three decades. According to Somit and Peterson, the continued persistence of the rational choice theory is not surprising because it "… fit comfortably into political science's 'political behavior is learned behavior' paradigm; it did not raise doubts about a considerable body of research in that discipline; its origins in economics endowed it with impressive scientific credentials ab initio; and graduate training at highly and less-prestigious institutions alike was quickly forthcoming" (1999, p. 43). Other reasons for the relative neglect of biopolitics research for many years might be the continued taint of social Darwinism and the eugenics movement in the United States (Dryzek & Schlosberg, 1995) or that many scholars had a professional stake in the viewpoint that all behavior is learned and, thus, not constrained in any way by biology (Somit & Watts, 1994). Furthermore, the experimental research required by a biologically based social science raises potential ethical problems that are not as conspicuous in noninvasive verbal response surveys. In other words, it is harder.

Given the initial inhospitable reaction of mainstream political science, and after considerable deliberation, the decision was made to organize biopolitics as a subfield under the rubric of politics and the life sciences. As noted above, establishment of the APLS with its own journal and annual meeting, although signs of maturation, might have further isolated biopolitics and made it even less visible to the discipline as well as to a new breed of behavioral researchers who were interested in genetics. In hindsight, treating biopolitics as a new subfield might have been a miscalculation. It should not have been treated as a separate area standing alone in isolation from the established subfields, because, while its relevance varies, biological concepts and variables have much to contribute to every area of political science, from international relations and comparative politics to political theory and public administration. Given the wealth of understanding about human behavior flowing from the life sciences over the past decades and its implications for the study of leadership, decision making, organizational behavior, and conceptions of our human nature, it has significance in public law and legislative and judicial behavior as well.

Somit and Peterson view the failure to move from organizational success to recognition in mainstream political science as having many reasons, including the intransigence of the discipline to a biological paradigm of behavior and a lack of a prestigious institutional base (1998). There also continues to be some second guessing about the wisdom of some of the strategies employed by the Association and the *Journal*. Goetze (2011), for instance, argues that by including biopolicy, the net was, in effect, cast too wide. Instead of trying to include all dimensions, the agenda would better have been concentrated on biobehavior. However, Johnson (2011) and others (Blank, 2011) dispute this and contend that the policy focus was an integral component and played a central role in establishing the Journal of Politics and the Life Sciences (*JPLS*) as a vehicle for meeting the APLS goal of advancing a more inclusive interdisciplinary framework. Another contributing factor might be timing. Although much of the early work in biopolitics was aimed at inculcating a biopolitical perspective in political science and initial efforts were made to publish in mainstream journals, the unresponsiveness of these journals to biopolitics submissions compelled those writing in biopolitics to seek more hospitable outlets for their work. Also, given the indifferent or antagonistic reactions to their appeals, a case for a more biologically oriented discipline emphasis in the biopolitics literature gradually shifted toward more substantive work.

ENTERING THE MAINSTREAM

The struggles faced by proponents of scientific biopolitics over the past four decades are undeniable. Recently, however, the study of the contributions of the life sciences to political science has changed from one of isolation to one of subdued acceptance. As noted earlier, much of the early work in biopolitics was an attempt to convince political scientists of the importance of a broad Darwinian approach to politics and for the need of a paradigm shift away from the SSSM. Although it seemed that this message was largely falling on deaf ears, this marginalization of biopolitics has begun to weaken. There are indications that it is gradually moving toward the mainstream triggered by an increased acceptance of more empirically based biopolitical research. Johnson (2011) finds it ironic that while the expressed purpose of the founders of the APLS was to integrate biopolitics into mainstream political science, it is beginning to happen, but not under the auspices of the APLS. For instance, in 1986, Douglas Madsen published a cutting-edge study of the role of serotonin in dominance behavior. Concomitantly, using the classic twin design, Martin and colleagues (1986) found evidence of a genetic basis for political beliefs. Although both studies remained practically unknown among political scientists, in 2005, Alford et al. published an acclaimed article in the *American Political Science Review* that reanalyzed the earlier twin data and expanded the coverage of ideology to include an analysis of the heritability of party identification.

Further indications that the biopolitical approach is progressively gaining a position in mainstream political science are unmistakable. For example, major political science conferences are hosting panels on biopolitics, thereby providing an expanding platform for the demonstration of innovative biopolitical work. Even more important, a recent content analysis showed that since the year 2000 the yearly number of journal articles and academic books in which the terms "evolutionary psychology" and "political science" co-occur has mushroomed, moving from practically zero to several hundreds (Petersen & Kennair, 2009). Moreover, Blank and Petersen (2013), in their study of the content of the three leading general-readership journals in political science, the *American Political Science Review*, *American Journal of Politics*, and *Journal of Politics*, found promising signs. After a low but highly fluctuating number of biopolitics articles annually between 1980 and 2004, since then there has been a steady increase in the number of scientific biopolitics articles in these elite journals. They also found that when political researchers today speak of biology they do not just say so incidentally but explicitly advance a biologically informed perspective on politics.

If the above is taken as an increasing, although still circumscribed, recognition of biopolitical research, the question naturally arises: why? While the potential reasons are numerous, three specific developments seem most decisive. First, and perhaps the most important reason, is simply a matter of timing. In general, science in the 21st century has an intensified orientation toward the biological level. The changing status of biopolitics, then, is part of a more general pattern in academia, where biological explanations of social phenomena are increasingly viewed as acceptable and even necessary. Two examples include the massive international activity flowing from the Human Genome Project beginning in the 1980s and the U.S. government initiative of the Decade of the Brain in the 1990s (followed by EU Human Brain Project and US Brain Initiative in 2013), which placed emphasis on expanded emphasis and funding for research on the structure and function of the brain. Another illustration is the awarding of the Nobel Prize in economics on many recent occasions to researchers who explicitly eschew the classical rational choice models and incorporate insights from the cognitive and biological sciences. As the rational choice theory becomes suspect even to many economists, it is likely that political scientists will eventually follow suit. Moreover, in society at large as well as in academia there is an increasing acceptance of biological explanations of phenomena that were viewed previously as exclusively within the cultural domain such as art (Dutton, 2010).

A connected tendency (and perhaps even a causal factor in fostering this acceptance) is that researchers from outside the social sciences have begun to bring their theories and methods to bear on classical social and political science matters. Political science topics are now legitimate, and often controversial, research topics in life science journals. One example is a study by Stanton, Beehner, Saini, Kuhn, and LaBar (2009) that investigated how election outcomes modulate testosterone levels in partisans. Simultaneously, political researchers are beginning to publish in life science journals. A group of political scientists, for example, published a study in *Science* demonstrating important physiological differences between liberals and conservatives in their reactions to anxiety-provoking stimuli (Oxley et al., 2008). Concurrently, life science researchers are beginning to publish in political science journals, often with political scientists as coauthors (e.g., Mattes et al., 2010). Such cross-fertilization is a significant sign that what we are experiencing in the political sciences is part of a more general trend in academia toward an acceptance of the biological components of social phenomena.

The second reason is methodological. This change has been stimulated, in part, by developments in human genetics and neuroscience, which facilitate the incorporation of biological variables into studies of social behavior. Evolutionary biology itself is, quite simply, hard to understand. As Richard Dawkins puts it, "it is almost as if the human brain were specifically designed to misunderstand Darwinism" (1986, p. xv). Accepting that complex human behaviors are under the influence of ancestral selection pressures requires triangulating evidence from a large number of disciplines such as psychology, archaology, and primatology and this, of course, requires knowledge about the research in these disciplines. However, changes in methodology have occurred that make the influence of biology on social and political behavior more discernable. Not without controversy (see Farah, 2014), the expansion of more affordable brain imaging techniques such as functional Magnetic Resonance Imaging (fMRI) has, for example, enabled researchers to directly demonstrate how specific social tasks draw on distinct neural circuitry, often operating outside conscious awareness, thereby undercutting the notion of the human mind as a general machine for making rational calculations (see, e.g., Ermer et al., 2006). Similarly, advanced approaches in behavioral genetics have enabled researchers to show the heritability of social and political orientations and potentially discover the specific genetic polymorphisms that influence these orientations (e.g., Fowler & Dawes, 2008). Consequently, accepting that biology matters now requires less cognitive effort because the methodological advances make it more directly demonstrable. This also lends legitimacy to more evolutionary-oriented accounts that focus on the structure of the biological mechanisms under investigation. These and other biologically based methodologies are discussed in Chapter 4.

The third reason for the budding acceptance of biopolitical approaches centers on theory or, perhaps more precisely, on the levels of analysis. This recent more receptive environment has been facilitated by a theoretical move within the biopolitical literature itself. Many early writings in biopolitics were concerned with developing a comprehensive, all-inclusive framework for how to understand politics in the light of biological theory (e.g., Arnhart, 1994; Masters, 1990). Accepting these writings, therefore, involved accepting the whole framework and one which contradicted prevailing theories. In contrast, many of the more recent publications on biopolitics deal with lower level theories that more narrowly focus on a particular topic and bring empirical evidence to bear on it. Some research has, for example, been preoccupied with demonstrating how narrowly specific political elements such as ideology (Alford et al., 2005), actual voter turnout (Fowler et al., 2008), generalized trust (Sturgis et al., 2010), and even imagination (Petersen & Aarøe, 2013) are genetically heritable. Other research has used the observation from evolutionary psychology that the human mind evolved to work within the small-scale environment of our ancestors to argue that citizens' political attitudes on mass politics are governed by the factors relevant in this small-scale environment. This research has then proceeded to validate this hypothesis on specific topics such as attitudes about war (Sell et al., 2009) or about criminal justice (Petersen, 2010).

The key point here is not only the need for grounding arguments in data but also that the move to lower level theories, amply supported by these data, has enabled the acceptance of the underlying biopolitical agenda because it enables the audience to accept the agenda piece by piece instead of all at once. In this respect, it is also interesting to note that the most cited journal publications of earlier biopolitical researchers were the empirical ones that address narrowly specified questions (e.g., Sullivan & Masters, 1994). Of course, this move to lower range theories on human social behavior did not happen in a vacuum but hinged on the successful development of these theories in the biological disciplines. In this way, the maturation of neuroscience, evolutionary psychology, and behavioral genetics themselves have paved the way for the application of these theories to political phenomena.

FRAMING AN INTERACTIVE PARADIGM OF POLITICAL BEHAVIOR

Early biopolitics researchers argued that to understand politics, it is necessary to apply knowledge generated through the study of the natural world: we recognize that man is a part of – not apart from – that natural world. These proponents of biopolitics advocated the need for a theoretical "paradigm shift" in political science from traditional and behavioral approaches to a life-science-based, evolutionary framework, although the suggested strategies for achieving this goal have varied widely as have the core problems and preferred methods. This new paradigm stressed the evolutionary history of the human species as a source of knowledge, reflecting a continuity that exists from early hominids and man's closest primate relatives to the present. Its followers sought to understand the biological bases of behavior and to comprehend the theoretical basis of political theory as an interdisciplinary enterprise that is both empirical and normative in its scope.

From its start, then, biopolitical theory encompassed biobehavioral theory and evolutionary theory. The former focused on the implications of the interactional model of human behavior and relied on studies at the level of the individual and small groups. The latter drew upon general systems theory and the evolution of social systems and political subsystems as well as the study of the biocultural evolution of institutions and structures. As in any genuine academic endeavor, the biopolitics movement exhibited several distinct strains within early biopolitical theory. Some, such as Masters, Schubert, and Corning, attempted to create theoretical frameworks that are comprehensive and fully informed by the full range of contemporary evolutionary theory and contemporary physics. Others, such as Somit and Peterson, drew upon those theorists and on sociobiology and ethology to challenge assumptions about contemporary democratic theory and practice.

The dominant model for the study of political behavior is the SSSM (see Box 1 and Fig. 1), which emphasizes the influence of environmental variables on decision making. In this model, the emphasis is clearly in nurture through political socialization of decision makers and the dynamics of groups in decision making. Evolutionary psychologists Tooby and Cosmides (1992) argue that the SSSM, with its roots in the work of theorists such as Durkheim and Weber, has positioned the social sciences away from thinking about how our evolutionary heritage affects behavior and the prevailing view that preferences are fully socially derived is incomplete. As Glendon Schubert (1976) pointed out, the social psychology approach

Box 1. The Standard Social Science Model (SSSM).

The term SSSM was coined by Tooby and Cosmides (1992) to describe the blank slate, social constructionist, or cultural determinist perspectives that they claim is the dominant theoretical paradigm in the social sciences. Although critics contend the SSSM is a straw man used to create a "false dichotomy between a manifestly untenable view and their own" (Richardson, 2007, p. 176; Wallace, 2010), the SSSM does capture the key features of the prevailing political science model, particularly in its assumption that biology is relatively unimportant to understanding political behavior. Among other claims, the SSSM presumes that the process of cultural transmission from generation to generation is maintained through a socialization process that acts to make the child like the adult of his/her culture. In sum, the SSSM holds that the mind of the individual is almost entirely shaped by culture and that facts about a culture are socially constructed, not products of human nature. What is also important is what the SSSM denies or, more often, ignores: The model holds that in discussing culture, one can safely neglect a consideration of psychology as anything other than the nondescript black box of learning, which provides the capacity for culture. Learning "must be the explanation for any aspect of organized human life that varies from individual to individual and from group to group" (Tooby & Cosmides, 1992, p. 32).

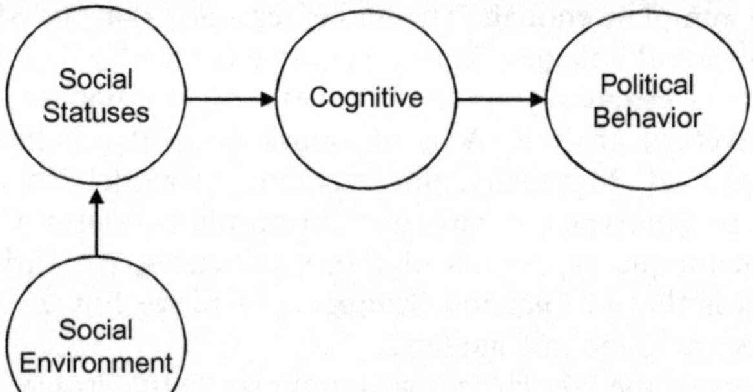

Fig. 1. The Social Science Paradigm of Political Behavior. *Source*: G. Schubert (1976, p. 177).

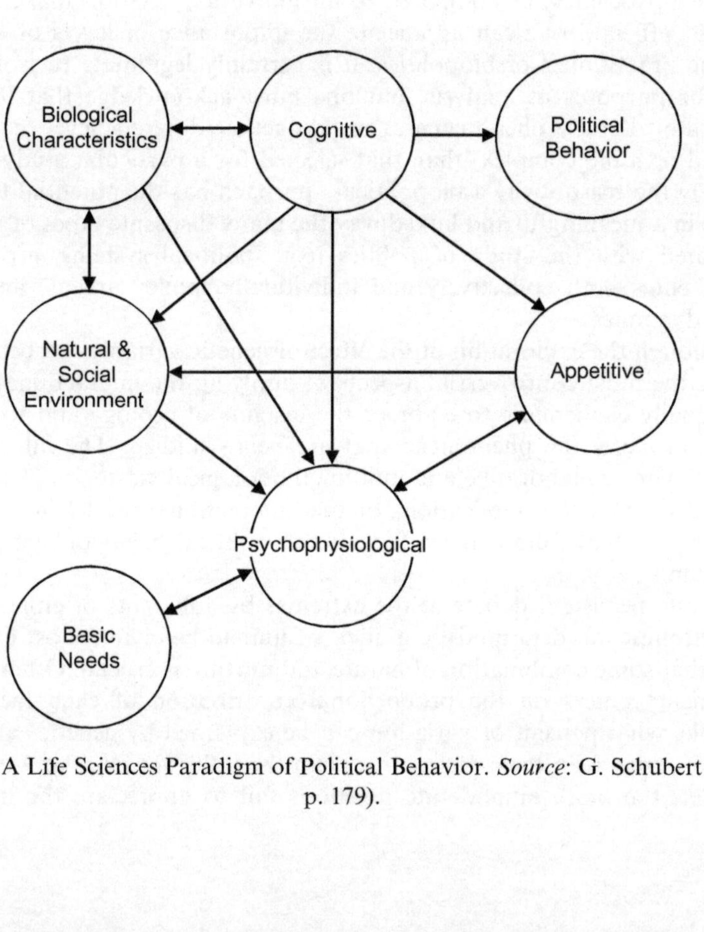

Fig. 2. A Life Sciences Paradigm of Political Behavior. *Source*: G. Schubert (1976, p. 179).

is simply not complete enough. The major reason is that the SSSM focuses on proximate causal linkages, ignoring completely the level of ultimate causation, which is essentially an evolutionary explanation that takes into account all levels of analysis. A more robust model that is interactional at its core is required. Minimally, the interactional model reflects the combined effects of genotype and environment on the organism with the result being the phenotypic expression of those influences. An early version of such a model is that of Glendon Schubert (1976, see Fig. 2). Other, more recent models are found in Chapter 2.

Practitioners of the life sciences well understand the problem of deciding on the level of analysis at which research is focused. The social sciences have tended to divide their interests, for example, in psychology and sociology, between the level of the individual and the level of the emergent group. When one considers the important differences between studying whole systems (e.g., societies) as compared to an individual decision maker (e.g., a political official), one can appreciate the importance of levels of analysis. For the practitioner of biopolitics, it is certainly legitimate to isolate one level for purposes of analysis, but one must acknowledge that it is also legitimate to study phenomena at an altogether different level (e.g., more reduced or more complex) than that selected for a particular study. This is precisely the reason why a biopolitical approach has the potential to incorporate in a meaningful and linked way the many disparate types of research associated with the study of politics (e.g., political systems, group processes, elites both collectively and individually, policy arenas, and small group dynamics).

Although the exploration of the effect of genetic variables on behavior is among the most controversial aspects of applying the interactional model, it is equally challenging to embrace the insights of ecology and to contextualize properly the phenomena that are being studied. The interactional model invites collaborative and multimethodological strategies of research. We will explore the implications of this interactional model in several of the chapters that follow as we focus upon political behavior and political evolution.

Despite persistent debate at the extremes by adherents of either genetic or environmental deterministic models of human behavior, most observers agree that some combination of nature and nurture is crucial. Often the disagreement centers on the proportional contribution of each factor, for example, what percent of variation can be explained by genetics and what by environment. In their desire to explain the influence in either–or terms, however, the more immoderate positions fail to appreciate the dynamic, interactive nature of the genetic/environment relationship. This neglect of biology becomes even more obvious when the brain is put into the equation. Very simply put, the interactive model holds that the genes and the environment are reciprocally related and therefore can influence one another over time. More importantly, both the genotype and the environment act to produce a specific phenotypic expression that defines the individual. Although this joint action serves to explain individual variation, it is not justifiable to generalize individual variation to population differences despite sexual, racial, and ethnic patterns in genotype and social environment.

The Brain Is Not a Black Box

While this basic interactive model places nature and nurture in perspective, however, it must be expanded to highlight the critical linkage factor, the brain. The more we understand the functioning of the brain, the more we are led to the conclusion that our behavior is fashioned by what our brain allows. This should not be interpreted as suggesting that our brains determine behavior, but rather that they affect or mediate every action and thought of both political leaders and citizens. Our capacity for enjoyment, suffering, and behavior is inscribed in neurons and synapses. As a result, our interpretation of the world, including the political dimensions, and our responses to it, depend on the internal organization of the brain. Therefore, in order to explain behavior or capabilities, one cannot minimize the role of the brain or view it simply as a black box or an empty organism as the SSSM tends to do.

As noted by Pierre Changeux: "The development of the neurosciences has brought another way of looking at behavior … The neuronal content of the black box can no longer be ignored. On the contrary, all forms of behavior mobilize distinct sets of nerve cells, and it is at their level that the final explanation of behavior must be sought" (1997, p. 97). Although it is unlikely that we will ever be able to link a particular behavior to specific neuronal activity, it is imperative that these neural dimensions are an integral part of any reputable model of political behavior. To ignore the role of the brain is no longer possible in light of what is now known even in the rather nascent stages of neuroscience. The incipient field of evolutionary psychology may help provide the tie between the biological bases of behavior and the larger process of evolution in the study of political, social, cultural, and economic behavior (For some examples of social scientists other than political scientists who are advocates of biosocial science, see de Waal, 1996; Eibl-Eibesfeldt, 1989; Walsh, 1995; Wright, 1994. For sources in bioeconomics, see a series of papers by Corning, 1995, 1996, 2000).

Elliott White offered visionary advice for political scientists in 1992 when he argued that new insights into neurobiology must be considered as an end to the notion of the empty organism because "the direct and significant impact of the brain on our emotions, intellect, and behavior has already been conclusively demonstrated in a number of ways" (1992, p. 35). For White, while a "systematic synthesis of the subject matters of the neurosciences and social sciences seems premature," clear evidence of the central role of the brain in accounting for human thought and action "dictates that this role be fully acknowledged within the social sciences" (1992, p. 19). What was premature in 1992, unquestionably, is now coming to fruition.

Similarly, in his critique of the rational choice theory and contemporary studies of voting behavior, John Wahlke asserted that they are flawed because of their reliance on a now discredited black-box model of behavioral psychology (1991, p. 179). According to Wahlke, this model is concerned only with functional relationships between observed stimuli and observed responses and not with "how" the connections are made so as to produce the responses. Inclusion of the brain provides this crucial dimension. Under a more integrated interactive model, the brain becomes the key facilitator of both genetics and the environment for the individual, as reflected in indisputable incontrovertible evidence that brain chemistry is critical to our understanding of behavioral patterns, personality, and a range of individual capabilities. This model offers a focus for analyzing the rich combination of genetic, environmental, and ultimately neural factors that define who we are. As well summarized by Rose McDermott:

> If we genuinely want to learn about behavior and the biological substrates of behavior, we need to mix levels of analysis and disciplinary fields precisely because of the limitations in vision inherent in each. … Any comprehensive model designed to estimate and examine this process must incorporate the reciprocal interaction between neural and cultural processes, accepting that the central dynamics involved remain behavioral as well as cognitive in nature. (McDermott, 2011a, p. 100)

Conversely, any study of the brain's structure, function, and role without consideration of the genetic bases and environmental influences that shape its attributes would be remiss. The brain requires constant stimulation by the environment in order to develop, and without sensory input and intellectual challenges of a positive nature, its full potential cannot be reached. Studies of infants demonstrate that while the plasticity of the developing brain is remarkable, without both suitable genetic grounding *and* environmental stimulation its growth will be constrained. Thus, while the brain is the great mediator it, too, is dependent on the genotype of the individual and the environment within which the individual operates.

Epigenetics: Complicating the Model

Although once dismissed as Lamarckian myth, epigenetic modification has become an important interface between the environment and the genotype. Epigenetics is the study of how the environment can affect the genome not only of an individual but also the development of one's descendants without altering the DNA sequence. Genes are transcribed to produce RNA and proteins, but they are not self-activating and are turned on and off by the epigenome. Although epigenetic regulation of the DNA could be stable throughout one's lifetime, it can also change in response to cellular signals from the environment that can result in long-term modifications with phenotypic consequences. Mitotic epigenetic inheritance occurs when somatic cells inherit the epigenetic marking of their predecessors and transmit them to subsequent generations of cells. While this process can bring about phenotypic changes in the individual, it cannot be passed to progeny because it does not affect the germ cells. In contrast, meiotic epigenetic inheritance involves changes to the DNA in the germ cells and, thus, can be passed from one generation to the next (Charney, 2011).

Researchers, for instance, have found that environmental factors such as abuse and trauma can trigger genetic changes by leaving traces on DNA in the form of epigenetic markers. Moreover, the evolution of acquired characteristics has been found in human populations that have experienced starvation, resulting in altered gene function in both the starved population and their offspring (Morris, 2012). In other words, experiences of earlier generations can change regulatory factors that affect gene expression, thereby substantially influencing the individual's physiology and behavior. As a result, particular environments may trigger entirely different cognitive or emotive processes and could be so formidable that they elicit a common response that leaves little room for genetic differences to manifest themselves (Hatemi & McDermott, 2012a, p. 10).

According to Crews (2008), it is now evident that cumulative experiences throughout one's life interact with genetic predispositions to shape the individual's behavior and that epigenetic traits can be passed down from one generation to the next. Dramatic events in past generations, therefore, could influence how an individual responds to events in their own life. For Crews, this new knowledge denotes a paradigm shift, one in which biology must be viewed through the lens of both genomics and epigenetics. Although most neo-Darwinians believe that epigenetic inheritance modifications cannot be inherited past one or two generations and are not an enduring basis for evolutionary change, the veracity of epigenetic inheritance is not in doubt (Morris, 2012). Despite debate over the significance of epigenetics to the evolutionary process, Hammock (2011a) suggests a need to increase our understanding of how the environment and the genome interact to establish the expression of one's genetic endowment, both within the individual's lifetime and across generations.

Epigenetics was unstated but integral to the NSF-funded Illinois Biology and Politics Summer Institute's focus on sociogenomics – the study of the genetic bases of social behavior and related environmental influences (Bucy, 2009). The Institute was organized around the theme that the notion of nature versus nurture is outdated, and that the relationship between an organism's genes and its environment is substantially more dynamic and complex than previously thought. The development of an organism is not determined solely by whether it has a particular gene but also by whether that gene is turned on and expressed. Moreover, the factors that determine whether a gene will be expressed are highly responsive to environmental stimuli. In particular, genomes can be responsive to stimuli associated with social behavior, which means that the same set of genes may result in divergent brain development and behavior depending on the social behavior to which the organism is exposed (Blanchard, 2009). Not only is the expression of genes often controlled by inputs from the environment, but also many genetic mechanisms require such inputs to function properly. As a result:

> Only by considering both the environmental and genetic sources of individual differences can we gain a deep understanding of behavior. The more we learn about how genes lead us into environments, affect our interpretations of the exogenous environments we encounter, and how our social environments may change our genetic expression, the more we can contribute to the discipline at large about which environments matter and why. (Hatemi, Dawes, Frost-Keller, Settle, & Verhulst, 2011, p. 81)

The last decade of scholarship in the political sciences has begun to integrate these two approaches into a more unified exploration of political behavior (Hatemi et al., 2011). Rather than treating social and genetic and neurobiological theories as competing models, the aim is to treat them as complementary approaches and to include elements of genes, the brain, and the environment into a unified theoretical approach that allows the possibility to further control and refine the nature of the intersecting relationships between biology and environmental stimuli and relevant political behaviors and preferences. "Only by considering both the environmental and genetic sources of individual differences can we gain a deep understanding of behavior. The more we learn about how genes lead us into environments, affect our interpretations of the exogenous environments we encounter, and how our social environments may change our genetic expression, the more we can contribute to the discipline at large about which environments matter and why" (Hatemi et al., 2011, p. 81). For instance, Hatemi (2013) found that both genetic and environmental variance on economic policy attitudes are moderated by financial risks. However, after 2 years, the phenotypic effects of the life events remain on most attitudes, but changes in the sources of individual differences do not. He concludes that genetic influences add stability while environment cues change.

Freese (2011) contends that evidence of the importance of genetic causes of social science outcomes and their potential for revision and elaboration of our understanding of social causes is abundant. Thus, it is crucial that social scientists integrate genetic causes into social science theories. The failure to do this can lead to large biases in social science studies of individual behavior. Given the habitual failure of conventional social science models to explain much of the existing variation in behavior, gene–environment interactions could be an important reason why individuals with similar social backgrounds and measured experiences often behave differently. Hammock (2011b) adds that we soon should see more examples of gene variation linked to behavioral variation. Although we will not likely be able to predict with meaningful accuracy an individual's behavior based on his or her genome, we will have increasing power to assess the probability of given traits and social behaviors at group levels.

Similarly, Hatemi concludes that "the paradigm shift toward a combined social and neurobiological approach has already occurred, and there is enormous growth in cross-disciplinary research focusing on social traits using neurobiological methods and approaches" (2011, p. 128). As evidence, he cites copious examples of recent neurobiological explorations of social traits including cooperative behavior, empathy, trust, altruism, social hierarchy, bargaining, risk, affiliation, leadership, punishment, social organization, ideology, attitudes, voter behavior, multivariate genetic models of personality and attitudes, genome-wide explorations of ideology, differences in testosterone levels and political competition and aggression, and different physiological reactions to threat between those with liberal and conservative orientations. No longer is the question whether this research should be taking place. According to Hatemi (2011), if social scientists were given the research opportunities to explore studies using similar approaches, it could lead to a transformation of the disciplinary approaches to behavior across disciplines and subfields. Already, the NIH has initiated a new funding mechanism, the Basic Behavioral and Social Science Opportunity Network (OppNet), to expand the funding of basic behavioral and social sciences research related to health traits.

RATIONAL CHOICE THEORY

Closely linked to the SSSM issue, the other major challenge a biopolitics-oriented interactive paradigm poses for contemporary political science centers on the rational choice theory under which collective social outcomes are explained in terms of individual rational decisions, treating individual actors as if they intentionally select beneficial norms. Despite the heated debates it has engendered in other social sciences, in political science, the rational choice theory has a sizable following and, given its emphasis on the rationality of human behavior and the determinative role of the environment is essentially a variant of the SSSM (Somit & Peterson, 2013a, 2013b). In the tradition of John Locke, it presupposes that each human's brain enters the world as a tabula rasa, with socialization processes then shaping preferences in reference to a specific culture. The rational choice theory erroneously assumes that actors consciously choose from a set of action plans to optimize outcomes based on the perceived strategies of other actors. However, extensive evidence from neuroscience introduced in Chapters 3 and 5 demonstrates that much action occurs outside a conscious, contemplative decision-making process. The modern evolutionist argument is that a broad range of human behaviors, including emotions, expressions, aggression, and drives for status, have an evolutionary origin. Open behavioral programs interact with the social environment to produce a phenotypic trait. For instance, all humans have the capacity for language, but specific languages are spoken only by those exposed to them. Closed behavioral programs, however, are more unyielding in their expression. For example, the meaning of many facial expressions is universal. Therefore, manifestations of closed behavioral programs will be more or less identical across cultures, while open behavioral programs will vary in substance, but not in underlying purpose (Tingley, 2007).

As noted by Prindle (2012), rational choice or public choice approaches to politics translate human decision making into what amounts to a mechanical process. The maximization assumption, which lies at the heart of neoclassical economic methodology, is a nonhuman analog. When applied to the human activities of politics, the value of such an approach stands or falls on the extent to which human minds actually function in the rational mechanical manner posited. If they do, then the method would be legitimized by the ability of the science to predict human behavior in both economics and politics (Prindle, 2012). As much scholarship over the past three decades has demonstrated, however, the rational choice approach fails to allow investigators to predict human behavior except in a very restricted subset of cases.

The threat of this revised paradigm to the dominant theories in political science is evident. A life-science-based political science requires the study of actual behavior, not merely attitudes and opinions, and it rejects the truncated account of human nature that a rational choice/economic man model relies upon for explanation. As cogently stated by Rebecca Hannagan:

> The suggestion that our political attitudes come from something beyond parental socialization, or culture, or social construct, or innate rationality is controversial because it conjures up fears of biological determinism and even social Darwinism. Placing something as seemingly esoteric as political attitudes in the realm of biology is frightening to some, threatening to others, and yet to others it is merely common sense, quite obvious, or even exciting. (2011, pp. 153–154)

Certainly, sociobiological theory subsumes rational choice and extends the calculation of self-interest to include genetic self-interest, but ultimately explanation requires evidence of enhanced survival potential if we are to understand why some political strategies are more successful than others over time. It is not a coincidence that the dominant perspectives enshrine the enlightenment ideal of the rational actor making individual choices and are, thus, consistent with political ideologies that exalt the autonomous individual. This individualistic model, however, ignores the complexity of collective action and the policy implications of our behavior based on an understanding of the dynamic interplay among genes, brains, and the environment. The behavior of the *political animal* and the implications for policy choices can be understood only within the context of a more sophisticated life-science-based model.

If we accept the strict Kuhnian usage of the concept of paradigm, then it is clear that biopolitics does not constitute a paradigm, for Kuhn has argued that the social sciences are "preparadigmatic" in as much as they do not have an established theoretical framework complete with all of the components associated with traditionally understood paradigms in the physical sciences. However, to the extent that evolutionary theory is a paradigm in the life sciences, then biopolitics may be said to possess something closer to a paradigm than some other theoretical frameworks extant in the social sciences (Hines, 1982a, 1982b). It must be noted, however, that among biologists, there is still disagreement as to what constitutes an agreed upon theory of evolution. For example, there are still significant disagreements among leading theorists of evolutionary biology as exemplified in the quarrel over the status of the concept of "punctuated equilibrium" in the theory of evolution (see Prindle, 2012). Nor have biologists come to a consensus as to the status of Wilson's sociobiology. Thus, in fairness, it must be said that evolutionary biology cannot function for biopolitics in the same way that "economic man" functions for the rational choice theory or that Skinnerian behaviorism functions for one version of behavioral political science. Nonetheless, we contend that an evolutionary theory of politics offers a potentially all-embracing model for the study of politics that comprises systems theory, cybernetics, rational choice theory, socialization theory, as well as cutting-edge developments in evolutionary psychology, psychobiology, and cognitive neuroscience. It also has the potential of providing a new naturalistic structure for political philosophy (see Arnhart, 1998; Masters, 1989).

BIOLOGY, HEALTH, AND BEHAVIOR

The genetic links to behavior are becoming increasingly evident, primarily as a result of innovative research from classic twin and candidate gene association studies. Chapter 4 examines these methodologies, while Chapter 5 discusses a wide range of relevant findings emanating from them. It is important to note that genetic factors have no direct influence on any behavioral traits, including political attitudes and behaviors. Rather, their influence is channeled through a complex pathway that starts with different genotypes (and their interactions with the environment or other genotypes), continuing through coding of proteins, amino acids that serve as signals in our body influencing our physiological functioning, our nervous system, and the operations of our brains. These genetically coded signals are subject to great variations even within genotypes, but, on average, they can influence dispositional traits across the genotypes which, in turn, can influence political attitudes and behaviors (Fazekas & Littvay, 2012).

Also largely neglected by political scientists has been the impact of health status on political behavior. How do disability and chronic illness, for example, affect political attitudes, interest, and participation? Even the normal process of aging can be a very significant individual and group variable in certain contexts (see discussion in Chapter 5 on age, age structure, and political decision making). Although many studies have found that health and illness can shape political activity, mainstream political analysis continues to focus exclusively on the same socioeconomic variables introduced in *The American Voter* in the 1960s. Moreover, while research has been conducted on the effects of aging on voting behavior and on the impact of health on leaders, the health factor as it affects the masses is seldom noted in mainstream literature. In addition to examining the relationship between health and political behavior, research is necessary to analyze the policy implications of unhealthy populations. For example, how might one's health status influence his or her position on health policy issues such as the Affordable Care Act or health care rationing?

A related area is the impact of illness on political decision making. Here, there has been interest among a few political scientists and life scientists who have found patterns that have troubling implications for all levels of political leadership. Studies by Robins and Rothschild (1981, 1988) on health, stress, and leadership behavior are the exceptions. In Chapter 5, we will examine what has been posited about disability and presidential decision making and about the effect of drugs and alcohol on political decisions. This research has importance not only in domestic politics but also for international politics but to date has largely been ignored by the discipline as a whole.

With regard to international relations, conflict must be viewed as arising not just because of environmental factors but also from varying innate predispositions on the part of the potential combatants (Long & Brecke, 2003). Although adopting a biological orientation does not require embracing a belief that all humans are inherently violent or peaceful, traits such as a proclivity toward violence vary dramatically from person to person, context to context, and gene pool to gene pool. As will be evidenced in Chapter 5, some people are more aggressive than others both in their personal lives and in their tendency to advocate collective violence against an out-group. Likewise, comparative politics could be greatly assisted by recognition of the fact that groups of people are genetically distinct from other groups of people. If understanding of politics is to advance, the grip of environmental determinism must be loosened, and appreciation of the role of biology must extend beyond think pieces and "calls to arms." Serious empirical research on the interaction of biological and environmental independent variables needs to become the order of the day (Alford & Hibbing, 2008).

THE LIFE SCIENCES AND PUBLIC POLICY

Another area of scientific biopolitics is research investigating policy implications of findings in the life sciences that impact individuals, society, and the global community. Issues of biopolicy concerned with individual persons include, among others, human genetic intervention, prenatal diagnosis, reproduction assisting technologies, brain intervention, nutrition, and death-related issues such as assisted suicide and definitions of death. Issues of biopolicy oriented toward society as a whole address matters such as genetically engineered organisms, public health, aging populations, biohazards, and sex differences. Finally, globally oriented biopolicy concerns include ecological and environmental issues, communicable diseases, population issues, and biological weapons. These areas represent the surface of an extensive array of biopolicy concerns that should be of interest to political science. Political scientists with an understanding of the life sciences are better positioned to provide valuable insights to policy makers as well as guidance to scientists as they deal with governmental agencies. Chapter 6 focuses on the former two categories, while Chapter 7 places emphasis on global issues tied to survival of the species.

Interestingly, Hatemi and McDermott (2012a) suggest that major policy issues in modern societies comprise basically the same concerns of reproduction and survival that confronted group life in ancient times and involve largely unchanged interpersonal traits. For example, current issues surrounding immigration are similar to the primal need to recognize and deal with out-groups, while foreign policy and punishment are matters of safeguarding the in-group. Likewise, health and welfare issues are essentially questions of the best way to allocate societal resources. Although the characterizations of issues and policies change across time and cultures, the underlying connections among the core issues for humans are enduring. Although the manifestation of genetic influence on these preferences becomes more complicated in large-scale societies and modern social structures, "genetic influences on attitude differences may be a remnant of ancient behavioral adaptation predating modern human society" (Eaves & Hatemi, 2008).

The implications of the life sciences for public policy are substantial in two distinct ways: the content of public policy and the process of policy making (see Caldwell, 1987, for an overview). Many of the most contentious policy issues currently are at their base biological, and this prevalence is bound to escalate in the future. With rare exceptions, to date political scientists have found themselves largely left out of the societal debate over these issues. Their absence from national commissions and other policy forums as compared to ethicists, sociologists, and particularly life scientists is striking. In large part, this reflects the lack of training and appreciation of life science perspectives by political scientists, some of whom may have entered the discipline to escape the worlds of science and mathematics.

This marginalization of political science from many critical policy issues of the day means that the valuable political perspectives that it could offer are instead given by life scientists, themselves, who often are not equipped to appreciate the nuances of the policy process. Moreover, students with interests in the policies related to these important issues often find themselves drawn to interdisciplinary programs in public health, environmental studies, and health administration rather than to political science departments, which offer them little in the way of coverage of these topics. Furthermore, programs in policy sciences and policy studies are frequently anchored in economics, not political science.

This raises a second problem in contemporary policy analysis, the question of how the policy process works. The interactive model discussed above directly challenges the conventional models of policy making. Over the last half century, many models have been promulgated, most prominently variations of the rational and incremental models where the debate has focused over how logical and systematic the policy-making process actually is. Alternately, bureaucratic and organizational models have been proposed that highlight, respectively, the importance of impact of bargaining among personnel and agencies as they pursue their own perceived interests and the impact of values and regular patterns of behavior that are common to any large organization. While still other models have placed greater emphasis on beliefs and ideology and begin to address the degree to which decision making is structured by social and political values that are not rational or impartial, notably absent in all these policy-making models are biological factors or any reference to the brain. Even belief system models that explicitly challenge rational models by suggesting that decision makers are not rational, rigorous, and objective agents, and that decisions are shaped by often unconscious perceptions and concepts, generally fail to account for the evolutionary dimensions of these preconceptions and beliefs (see Box 2). In other words, they too neglect the biological foundations of decision making and the intricate contributions of both emotion and cognition. Chapter 5 examines the implications of emerging research in genetics and the brain on our understanding of the policy-making process.

Rationality has come to mean the conscious, goal-oriented reasoned process by which an individual, expressing and thus also revealing his or her preferences, chooses a utility-maximizing action from among an array of alternative actions. Although the rational choice theory does not specify that people must be "conscious" or fully in "control" of their decision-making process in order to be rational, it is most often used in situations where the posited actors are assumed to be conscious and acting intentionally (Tingley, 2006, p. 27). Also, while a few cognitive theoreticians have considered a role for emotion in decision making and some organization theories and agent-level theories allow for the incorporation of emotional variables, most either ignore it or see its role as secondary or marginal and view its formal consideration as counterproductive.

The heightened understanding of the role of human biology in decision making must be better integrated into policy analysis and into our

Box 2. Social Constructivism.

Yet another approach, social constructivism, explains phenomena by reference to social interactions rather than anything innate to human beings and claims that society and societal phenomena can be understood solely as the product of social interaction. Thus, emotion and behavior associated with it are merely social constructs that are learned and reinforced through social discourse. Long and Brecke (2003), however, argue that emotion is not reducible to a social–cultural construct, as constructivism assumes. Certainly, our emotion is provoked by social–cultural circumstances and in part expressed according to social–cultural conventions, but in between provocation and expression is nothing but the brain and the body. Although the human mind is capable of recognizing and describing subtle emotional states that reflect one's particular society and culture, basic emotions such as fear, anger, sadness, and joy transcend culture and represent universals.

models of the policy process. For instance, Graham Allison's introduction of the organizational process and governmental (bureaucratic) politics models as alternatives or supplements to the rational actor model was a significant improvement of our understanding of political decision making. However, four decades later, it is just as important that our heightened knowledge of the biology of human behavior and the evolutionary bases of organizational behavior be incorporated into more inclusive models of policy making. As in the study of political behavior, the context of policy making must be recast in light of new findings in the neuroscience, genetic, and evolutionary spheres. For too long, political science has ignored the evidence from the life sciences as it affects policy content and process, despite the warnings of the discipline's founders that biology must have a central place. This imbalance must be redressed by the development of a revised paradigm, which gives full appreciation to human biology.

ORGANIZATION OF THE BOOK

Chapter 2 analyses the vast implications of biopolitics for political theory and the need to reevaluate basic assumptions of the prevailing political science paradigm. It reviews the contributions of the major figures in the development of biopolitical theory. One of the major problems to be faced in adopting a more biology-based approach to political science centers on methodological problems inherent in a more inclusive, holistic model. Chapter 3 discusses these difficulties, but demonstrates that they are surmountable. It addresses the issue of level of analysis at length and discusses the dangers of the dependence on any single methodology, such as the prevailing method of survey research, to explain behavior.

Two areas where biology has attracted considerable attention are comparative politics and international relations. Chapter 4 examines the contributions of biopolitics to these subfields and reiterates the need to recognize the evolutionary bases of human behavior and the necessity of more interactive, inclusive theories of ethnic conflict and recurrent wars based on cognitive neuroscience research. Chapter 5 shifts attention to the biological bases of behavior. It first summarizes the findings of biopolitical research over the past several decades on the impact of health and nutrition on political behavior. It also describes research on the biological foundations of elite behavior and leadership. It then examines in detail the current state of knowledge concerning the genetic and neurological bases of human behavior.

Chapter 6 shifts attention to a range of biopolicy issues facing individuals and society and the challenges they create for traditional policy-making institutions and shows how crucial a biological-science-based approach is in dealing with these complex policy problems. Chapter 7 expands this analysis by examining the global environmental and population policy arena. It argues for a synthesis of biopolitical theory with work on political ecology. Especially when combined with bioeconomics, this can present a formidable theoretical perspective for the articulation of a life science informed biopolicy for the environment.

In the concluding chapter, the case again is made for a paradigm shift toward an interactive model. It suggests that, rather than lead to biological determinism which has often been charged by opponents of biopolitics, this inclusive paradigm allows us to counter deterministic protestations more effectively than by denying any biological influences. The chapter concludes by reiterating the call for considerably more interdisciplinary research and a human-centered political science in which the empty organism is given life and the black box at the center of the SSSM is brought into the light.

CHAPTER 2

BIOLOGY AND POLITICAL THEORY

Joseph Losco

We are well into the second generation of scholarship exploring the relationship between biological knowledge and political theory. Whereas the first generation, led by a few dozen pioneers, advanced diverse theoretical pathways by which life science findings might be incorporated into the study of political life, the current generation is producing empirical findings that promise to fulfill the goals of their intellectual forebears. Both generations, however, have faced a skeptical mainstream that remains largely unconvinced about the relevance of their findings.

In this chapter, we will briefly review the emergence of first-generation scholarship that built the modern foundation of biopolitics and then proceed to examine second-generation literature, its connection (or lack thereof) to earlier work in the field, assess the impact of these works on mainstream political science, and suggest future directions. The focus here is on contributions to political theory, broadly construed to include both empirical (including epistemological issues and especially those related to paradigm change in the discipline) and normative issues. In a review of literature that has vastly expanded over the past four decades, it is impossible to cite, let alone explore, all of the contributions. As a result, the focus here will be on some of the more representative entries to the field.

Politics and the Life Sciences: The State of the Discipline
Research in Biopolitics, Volume 12, 35–65

ISSN: 2042-9940/doi:10.1108/S2042-994020140000012000

PIONEERS IN BIOPOLITICS: THE FIRST GENERATION

It is not surprising that much of the biopolitics literature produced by first-generation scholars was theoretical in nature and drew disproportionately from those who called themselves political theorists. When knowledge from adjacent fields of inquiry is imported into another, scholars are likely, at first, to speculate on how those findings might illuminate intractable findings within their own field. This is especially the case when one's field, like political theory, deals with perennial questions that have eluded easy resolution for millennia.

As noted in Chapter 1, modern biopolitics can be traced back to the mid-1960s when it quickly developed a loyal core of adherents who contributed early and often to first-generation literature. This era witnessed an explosion of important findings in the life sciences and an expansion of the methods by which the biological roots of behavior could be explored. Primatology, ethology, human behavioral ecology, behavioral genetics, evolutionary game theory, neurobiology, sociobiology, and related fields surged during this period with new discoveries about animal populations and, sometimes speculative, imperial incursions into the social sciences. In return, social scientists began paying attention to these findings and developing their own niche subfields. Anthropology had already a long-standing branch of biological inquiry, but biological thinking soon found its way into economics (see Zak & Denzau, 2001) and psychology (Siverman & Fisher, 2001). Lynton Caldwell (1964), Albert Somit (1972), and Glendon Schubert (1976) signaled political science's entry into the conversation. Each of these pioneers had already been established scholars in the fields of public administration, political theory, and judicial behavior, respectively, and their interest signaled the potential widespread applicability of life science findings to the discipline. At the same time, Edward O. Wilson's *Sociobiology: The New Synthesis*, published in 1975, drew extraordinary interest among scholars in the social sciences, and political science was not least among them.

By 1982, the literature in the field had grown substantially and attempts were made to catalog its development. One review, ostensibly about the impact of sociobiology in political science, but including biological approaches like evolutionary theory, ethology, population biology, and neurobiology more broadly (Losco & Day Baird, 1982) found that from 1975 to 1980, 88 papers and publications had been produced in the

English language, representing the work of 36 authors. However, a striking feature of this productivity was the fact that a small subset of these authors produced the bulk of the literature. Of the 36 contributors during this period, nine produced four or more entries, three produced two or three entries, and 24 produced just one entry each in the field. It was determined that, especially among core contributors, the largest proportion of respondents self-identified as scholars in political theory or philosophy. As a result, much of the literature in the early period of the development of biopolitics produced a wealth of ideas about the ways in which life science findings might enrich our understanding of the perennial concerns of political philosophy. Additionally, it covered epistemological issues arising from the mingling of these fields. A sample of the theoretical contributions by some of the more prolific first-generation authors are presented here.

Empirical Biopolitical Theory: Paradigm Change

> Though … excursion into the implications of evolutionary biology and ethology has been sketchy, it should show that the utility of these biobehavioral sciences lies first in their forcefully directing the attention of political researchers back to broad and fundamental problems, most of which have been central concerns of traditional political philosophy and political science in earlier times. The biobehavioral perspective views political phenomena from the standpoint of its evolutionary history. It thus provides the only justifiable baseline for comparing variations in the behavior of various human groups. Second, it clearly forces us to reexamine our two-dimensional, oversimplified, supercognitive, social-psychological model of the acting individual. It makes us take into account important uniformities in all human behavior so far ignored by political behavior research. (Wahlke, 1979, p. 27)

These words penned by a former President of the APSA reflected both the dissatisfaction of some scholars with the standard methods employed in the study of political science and great optimism about the findings emerging from the burgeoning life sciences. Indeed, John Wahlke criticized political science research for its failure to concentrate on meaningful macro-level questions that are theoretically driven and for its reliance on subjective measures and self-reports rather than on the actual behavior of human beings. By contrast, findings emerging from life science research, a new brand of biopolitical scholars, and several political science labs like Stony Brook gave him hope that his profession would move past its prebehaviorist past and toward becoming a biobehavioral science.

The same refrain has been heard many times and in many ways since Wahlke's address. The SSSM, relying as it does on the manipulation of environmental variables, has been assailed for imposing an environmental or cultural determinism (Corning, 1971), for advancing a false blank slate view of human nature (Pinker, 2002) and for assuming the human organism to be nothing but a black box (Hopple, 1980) or an empty organism (White, 1992). As a result, a common theme of the proponents of biopolitics was the need for a paradigm shift to address these shortcomings.

Dissatisfaction with the direction of the discipline and rapid advances in the life sciences made grand assertions by those outside political science seem plausible. Particularly noteworthy was Wilson's contention that sociobiology would "cannibalize" the social sciences (1975, p. 575). At its core, sociobiology posits the evolution of adaptive behaviors and strategies via inclusive fitness. Accordingly, individuals can achieve reproductive success not only through their own reproductive activities but also by advancing the reproductive fitness of related individuals with whom they share genetic material. This discovery allowed biologists to explain a number of social behaviors like reciprocity and altruism that had previously eluded explication through strict Darwinian individual selection. Inclusive fitness was given a strong boost by the works of Hamilton (1964), Trivers (1971), and others who demonstrated robust empirical support through experimentation and observation in the natural world. The evolution of acts like altruism could now be explained by reference to the fitness benefits conferred upon related others. Thus, a warning cry issued by a prairie dog could save a sufficient number of kin to compensate for the loss of its own genes at the hands of a predator alerted by its call. Sociobiologists also showed that species can alter the sex ratio of their offspring depending on social rank and that levels of parental investment necessary to raise offspring to reproductive age can account for differing mating strategies by males and females (Trivers & Willard, 1973).

The potential of sociobiology, and evolutionary theory more broadly, for elucidating human political behavior led a number of political scientists to apply this new perspective to the study of phenomena like aggression and warfare (Falger, 1989; van der Dennen, 1995), leadership (Schubert, 1988), sexual politics (Schubert, 1989), reciprocity and coalition building (Axelrod, 1984), and authoritarianism (Somit & Peterson, 1997).

But reaction to sociobiology's more extreme claims led even its adherents to reexamine its imperialistic desires. By 1977, Wilson, himself, proposed a more modest role for his intellectual offspring: "Biology is the key to human nature, and social scientists cannot afford to ignore its emerging principles. But the social sciences are potentially far richer in content. Eventually they will absorb the relevant ideas of biology and go on to beggar them by comparison" (1977, p. 138). Instead of the prospect of social sciences becoming a branch of sociobiology, he now proposed the goal of consilience, the progressive unification of all strands of knowledge under the umbrella of evolutionary theory for the indefinite betterment of the human condition (1998). Only when social scientific explanations of cultural, economic, and political behaviors are carried back to their causal bases in biology our understanding will be scientific and the false boundary separating the various sciences vanish.

Wilson's revised vision of knowledge integration is closer to what some political scientists had in mind. Glendon Schubert's reading of paradigm change accommodates both evolutionary and revolutionary elements. "In both physics and biology," he claims, "the shift from the classical to the modern paradigm has been ... evolutionary. The revolutionary change came when classical physics and classical biology became differentiated from antiquarian natural science. Both the theory and the substantive knowledge of the macroparadigms that prevailed before Newton and Darwin were displaced by *their classical* works. Then the resulting macroparadigms became synthesized into their respective neoclassical forms" (1989, p. 300). Both earlier and later forms of thinking may proceed together for some time and find continued applicability in their own spheres, "much as Newtonian physics remains extremely useful and sufficiently correct as a predictor of macroscopic physical events on or near the surface of the earth – although no round trips to the moon could have been accomplished in exclusive reliance on classical physics" as opposed to special relativity theory. But paradigm change is sometimes accelerated by a "change in the *major* emphases at the cutting edge of scientific inquiry" (1989, p. 299). Political science had long borrowed from other fields using diverse methods, producing parallel lines of investigation. While these lines of research may have been informative, the field as a whole lacked unity. Schubert's hope, as well as those in the biopolitics subfield generally, was that evolutionary theory would serve as the unifying thread that would integrate existing disciplinary findings and change forever the discipline's future direction, ultimately aligning it with the natural sciences.

The Nature of Politics and the State

With evolutionary theory as an umbrella, Schubert and others proposed complex models as guides toward a unified theory in the field. At the heart of their enterprise is the desire to locate human politics within the evolutionary history of the species and thereby understand the biological underpinnings of political behavior and the state. Schubert advanced a multilevel model which begins with basic survival needs that can find satisfaction through the organism's interaction with the material and the social environment. Both the needs themselves and their satisfaction are conditioned upon individual psychophysiological factors like health, sex, and age. Appetitive behaviors issuing from our needs direct us in searching the natural and social environments and the search is further guided by feedback which might not only satisfy our needs but also further stimulate our appetitive senses (1989, p. 29). The model, informed by Schubert's own ethological investigations, facilitates cross-species comparison in exploring how needs are met. Such comparisons may be useful in understanding how behaviors evolved and in recognizing the contexts in which evolved desire-satisfaction strategies are effective. However, Schubert does not mean to imply that behaviors across species are of common genetic origin; merely that they may serve similar fitness functions. He goes to pains to discuss why an understanding of the unique properties of the human brain is essential for an understanding of the distinctive nature of human politics.

An alternate model is advanced by Roger Masters (1989) who attempts to consolidate a wide range of research in the life sciences. In *The Nature of Politics*, Masters argues that the dominant social scientific paradigm based on Newtonian physics had been rendered obsolete in the face of the far more powerful and relevant paradigm of evolutionary biology. When politics is considered within the context of the evolutionary processes from which it emerged, Masters argues that, we can bring considerable insight into our understanding of the function of the state, the operation of bureaucracies, and the dynamics of political behavior and leadership. Within the framework of evolutionary thinking, we are told, political theorists will be in a better position to evaluate claims and debates between materialists and idealists, moderns and ancients, and nativists and environmentalists.

Human behavior, according to Masters, is nested within physical and social systems and irreducibly involves interaction among three levels of operation: species, society, and individual (see Fig. 1). "Human behavior is the product of an integration within the brain and central nervous system

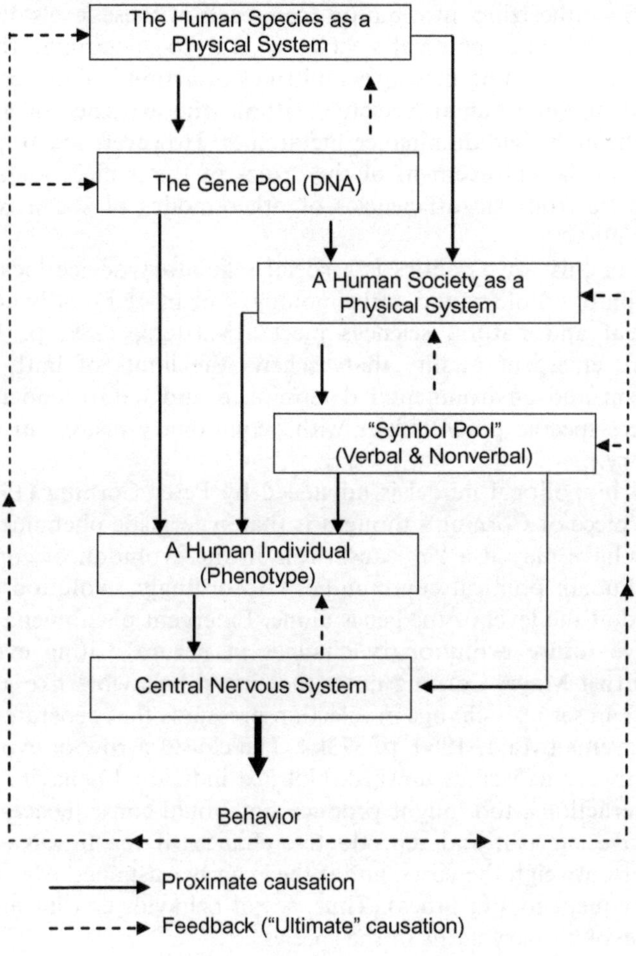

Fig. 1. An Interactional [Evolutionary] Model of Political Behavior. *Source*: Masters (1989, p. 134).

of phylogenetically selected information transmitted by the genes, historically selected information systems transmitted by language and cultural symbols, and individually learned information acquired during the life cycle" (1989, p. 135). More specifically, *political* behavior is that part of the human repertoire involving the attributes of bonding, dominance, and submission and is concerned with the regulation of social life. Because systems involved in synthesizing information from each of these levels are never in complete equilibrium, political systems emerge to "determine the priority of potentially conflicting messages and rules of action." This function may be satisfied in some human societies without the existence of a state, for example, through rigid dominance hierarchies. However, social institutions specialized in the enforcement of the "rules of the game" – states – are likely to arise from "insufficiencies of other modes of social regulation" (1989, p. 139).

Viewed in this way, politics is a social regulatory device located at the "intersection of ethology and anthropology – or more broadly at the point where social and natural sciences meet" (Masters, 1989, p. 140). This assumes an emergent quality that eschews the limits of both biological reductionism and environmental determinism and nature and nurture. It links species-specific potentialities with evolutionary history in telling the human story.

A third biopolitical model is advanced by Peter Corning (1983, 2005). The centerpiece of Corning's thought is that synergistic phenomena of various kinds have played a key causal role in the evolution of complex systems like human political communities. Accordingly, evolution cannot be understood at the level of the genes alone. Emergent phenomena may arise which drive future evolutionary changes in nature. Citing evolutionary biologist Ernst Mayr, Corning notes that novel behaviors like a shift to a new niche can set up a change in selection pressures that generates new evolutionary events (Mayr, 1960, p. 373ff). The causal arrow of evolution can point downward as well as upward. Not just individual behavior but cooperative interactions, too, might produce functional consequences in nature that improve survival and reproductive chances if the fitness advantages they confer outweigh the costs, and if they can be sustained over time without a subsequent loss in fitness. Thus, social behavior can have an impact on future evolutionary paths of the species.

Applying this conceptualization to politics, Corning defines a political system "as the cybernetic aspect, or 'subsystem' of any socially organized, cooperating group or population. Politics in these terms refers to social processes that involve efforts to create, or to acquire control over, a cybernetic social system, as well as the process of exercising control" (2004, p. 9). He goes on to note that control or power comes in many forms, including seniority, inheritance, expertise, and, of course, lethal force. But power is a means, not an end. The end is reproductive fitness. Historically, fitness has been enhanced in complex human social communities by the increase in goods and services that the society produces and makes available to its individual members. Stability in these communities is heightened by sets of rules and hierarchies that enforce cooperation, giving rise to the state. Viewed from this perspective, the state is a superorganism producing *mutually beneficial* economic synergies, "an aggregation of interests into an immensely complex system based primarily on *mutualism and reciprocity*" [emphasis added] (2004, p. 46). In other words, it serves as a collective survival enterprise whose success depends on the continuing positive synergistic effects of the many units that make up the collective whole. Corning recognizes that while the state may have evolved as an instrumentality for "self-government" and the pursuit of the public interest, its overarching purpose is all too often subverted for individual purposes.

Albert Somit and Steve Peterson believe that this subversion may be the rule rather than the exception. While the state may have emerged to solve collective survival problems, as Corning suggests, it did so in the face of evolved human propensities for hierarchy and dominance. As a result, "the overwhelming majority of political societies have been authoritarian, that is, characterized by the rule of the few over the many, by command and obedience, and by dominance and submission" (1995, p. 19). Somit and Peterson draw support for their theory from the history of the nation state, ethological and primatological studies, ethology, small group psychology, and anthropological studies of primitive societies. Given the persistent similarities between primitive societies and our own, the authors conclude, "it is truly difficult to escape the conclusion that these cross-cultural consistencies of social structure reflect a genetically transmitted behavioral tendency" (1995, p. 27). Of course, democracies do exist but only when key supportive conditions emerge (see Box 1). These include a sound national educational system, dense communication networks, and positive economic conditions (see also Vanhannen, 1984).

Normative Biopolitics

Evolutionary thinking has long offered intriguing possibilities about unlocking insights into human nature that could inform normative political theory. For many normative theorists, human nature is the starting point for political discourse (e.g., Germino, 1975; Somit, 1981). Darwin himself speculated about the relevance of certain human attributes for moral theory. Human emotions like sympathy, he believed, combined with a social instinct, the ability to reason, and an early evolutionary history

Box 1. What Has Biopolitics Contributed to Political Science?

Somit (2011) contends that biopolitics helps answer the question as to why are democracies so rare and fragile. There is a general agreement that democracies require a unique combination of social, political, and economic prerequisites or enabling conditions. When met, democracy is at least possible; when not, a viable democracy is very, very unlikely. He suggests that from a neo-Darwinian perspective, these special conditions are necessary because democracy runs counter to a powerful human behavioral tendency. "Biologically speaking, humans are social primates and, over several million years, evolution has endowed the social primates with an innate predisposition for hierarchical social and political structures. That is, social primates almost invariably form groups, troops, tribes, and societies characterized by marked individual differences in terms of status, dominance and submission, command and obedience, and by unequal access to many of the good things of life. The primary reason for the prevalence of authoritarian governments, for the rarity of democracy, and for why democracy demands such special enabling conditions is to be found in our genes" (Somit, 2011, pp. 96–97).

spent living in small groups where cooperation was a requisite for survival to make a human a moral animal.

> At the moment of action, man will no doubt be apt to follow the strongest impulse; and though this may occasionally prompt him to the noblest deeds, it will more commonly lead him to gratify his own desires at the expense of other men. But after their gratification when past and weaker impressions are judged by the ever-enduring social instinct, and by his deep regard for the good opinion of his fellows, retribution will surely come. He will then feel remorse, repentance, regret, or shame; .... He will consequently resolve more or less firmly to act differently for the future; and this is conscience; for conscience looks backwards, and serves as a guide for the future. (1874, chapter 4)

Taking their cue from Darwin, first-generation biopolitical theorists moved to incorporate emerging life science findings within moral philosophy in order to sketch the outlines of an evolutionary ethic. One of the first to attempt this was Richard Shelly Hartigan (1995). According to Hartigan, "the two thousand year tradition of higher law/right should be viewed as an intuited connection between the notions of nature, human nature, morality and justice" (1995, p. 87). The historical quest for natural right is an expression of humankind's natural instinct to come to grips with the problems of survival in a social environment. A review of that history shows universal patterns and ideas which come to represent a common set of moral prescriptions which remarkably resemble parts of the Ten Commandments. In particular, Hartigan points to what he calls the "social commandments" (four through ten). "Filtered through various ancient and modern translations, and reduced to their bare essentials, the social commandments deal with the basic issues of life, lineage and property protection, mutually maintained through the assistance of truth telling. In varied formulations, these are the core principles of every human society even today. … Their intuitive binding force can be seen as the necessary ingredients in bonding, nurturing, and protecting as our species became Homo sapiens" (1995, p. 100). Though he died shortly after he penned these words, Hartigan was certain that continued progress in anthropology, ethology, and behavior genetics would support his conclusions.

The search for moral universals also informed the work of James Q. Wilson (1993). However, unlike Hartigan who believed that moral injunctions themselves may have an evolutionary pedigree, Wilson identified universalism with evolved human traits, most particularly with what he calls a universal "moral sense," which encompasses both a pervasive tendency to make moral judgments and a set of social habits that guide our social interactions. These derive, Wilson believes, from our shared evolutionary heritage:

> If Darwin and his followers are right, and I think they are, the moral sense must have had adaptive value; if it did not, natural selection would have worked against people who had such useless traits as sympathy, self-control, or a desire for fairness and in favor of those with the opposite tendencies (such as a capacity for ruthless predation, or a preference for immediate gratifications, or a disinclination to share). (Wilson, 1993, p. 23)

Wilson cites now classic research in kin selection and reciprocity as well as studies in primatology and neurobiology as evidence of our inherited predisposition toward morality. However, but he is not content to rest his case on biology alone. While those who say culture is everything are clearly wrong, he notes, so too are those who say culture is nothing. Our innate moral sense must be guided and nurtured to come to fruition, and, even then, it is in competition with other elements of our natural heritage.

"I agree," he writes, "that there is a universal human nature, but disagree that one can deduce from it more than a handful of rules or solutions to any but the most elemental (albeit vitally important) human problems." Some universal truths, such as man's sociability, coexist with other universal truths like human ambition. "The passions of man are in conflict; his moral sense is one of his calmer passions, but it cannot always prevail over its wider rivals and its reach is uncertain and contingent" (1993, p. 219). Wilson aligns his views with those of Hume and Adam Smith, but finds in modern biology more secure empirical footing.

Roger Masters (1990) extends the discussion of the import of biological findings for political and moral theory. He calls for a naturalistic ethic that acknowledges: (1) respect for individual and cultural differences; (2) virtue as a natural obligation to those benefiting from the advantages of civilized society; and (3) natural justice as a standard for ethical judgment. The first of these principles follows from the natural diversity of humankind where no two individuals (even identical twins) are exactly the same. Since differences are the raw materials for selection and since we can never know which of us may carry a gene variation that confers a fitness advantage, nature informs us we are all "equal" in a profound sense (1990, p. 205). According to the second principle, virtue is a rational foundation for ethical obligation within a civilized society. Reproductive success in the age of modern society depends on the moderation of our own selfish needs by cooperation and reciprocity. These same elements – cooperation and reciprocity – give rise to Master's third principle and constitute the essential ingredients of natural justice. While Masters admits that it is difficult to derive specific injunctions from this standpoint, it is possible to know injustice when one sees it. He suggests that arbitrary use of violence and genocide are clear examples of the latter. Ultimately, Masters believes that modern biology leads to a view of natural justice akin to that of Aristotle, albeit stripped of its prescientific moorings or the theological baggage associated with natural law.

Larry Arnhart (1995, 1998) also adopts an Aristotelian perspective in his biopolitical theory. Arguing that the ultimate aim of politics is "to form the character of human beings to promote some conception of the best life" (1998, p. 1), he attempts to show that the root of this endeavor lies in our evolutionary heritage. Evolutionary biology, he believes, can point us to our natural human ends and political life can lead us to fulfill these in a way that allows humans to flourish. He lists 20 natural ends which like Aristotelian natural right direct us toward our natural destinies. Among these desires are health, sexual identity, familial bonding, and several distinctive desires like the desire for a complete life, aesthetic pleasure, and religious understanding.

Human moral conduct is motivated by an innate moral sense which steers our social behavior. This sense is not entirely unique to our species. It stems from the interplay of "knowing, desiring, and evaluating" (Arnhart, 1998, p. 24). Driven by our natural inclinations in a social environment, moral conduct is evident in many species that have the capacity to survey the landscape for the best means to satisfy desires and then take those actions necessary to fulfill them. In humans, this is a more elaborate process involving, in the ideal case, prudence as our guiding principle. Arnhart identifies four stages in the development of human morality: (1) social instincts led human ancestors to feel sympathy for others in their group; (2) intellectual faculties led to the ability to discriminate and adjudicate among instinctive desires; (3) language allowed expression of moral judgments, including right and wrong; and (4) habit allowed individuals to conform to opinions about right and wrong. A possible fifth factor, tribal warfare, may have facilitated in-group cooperation and out-group conflict (1998, p. 76). Arnhart sees evolution as a complex multilevel process involving upward and downward causal relationships, and natural sympathy plays a key role in facilitating reciprocal relationships including justice.

Arnhart readily admits that his theory rests on a teleological understanding of nature: "the idea that nature acts for the sake of ends" (1998, p. 3). This is an understanding, he believes, that is perfectly in line with Darwinian thinking. In fact, the biological explanation of animal behavior requires teleological or functional concepts. "Animals act in a goal-directed manner to satisfy their natural needs based on the information about the changing environments in which they live" (1998, pp. 21–22). Arnhart notes the affinity between the Aristotelian term and Ernst Mayr's biological derivation "teleonomic" (1998, pp. 247–248).

Armed with this biopolitical theory, Arnhart explores several illustrations of how Darwinian natural right can be used to distinguish between social relationships that are in accordance with nature and those that are not. Familial bonding and sexual differentiation belong to the first category, slavery and female circumcision to the latter. The former are distinguished by their ability to fulfill the ends toward which human nature aims, the latter frustrate those aims.

In a later work, Arnhart (2010) proposes to show that biopolitical theory of the sort he has outlined in *Darwinian Natural Right* is the corrective to political science's shallow treatment of the subject matter. Echoing some of the concerns voiced by John Walhke, Arnhart takes modern political science to task for its failure to appreciate the role of history, morality, judgment, emotion, and religion in the political life of human beings. Instead, political science is fixated on a model of Newtonian physics that misses the intricacies and complexities of political life. "At best, political science can make general predictions about political patterns. But it cannot make specific predictions about political events" (Arnhart, 2010, p. 39). It can yield prediction without insight. By contrast, evolutionary biology and the social sciences are historical sciences of emergent complexity. A biologically informed social science grounds human action in the evolutionary history of animals, involving three kinds of historical narrative: natural history, cultural history, and biographical history.

> Although biopolitical science cannot provide deterministic laws, it can provide probabilistic regularities, which support falsifiable but fuzzy predictions of behavioral patterns. For example, evolutionary game theory has developed formal models of the natural behavioral repertoire … and those models can generate falsifiable predictions that can be tested through experimental game playing. (Arnhart, 2010, p. 39)

Indeed, game theory has played an important role in political science research in recent years and its contribution to biopolitical theory in particular is important. Using a version of the Prisoner's Dilemma in a series of iterated tournaments, Robert Axelrod (1984), for example, found that a strategy called "Tit for Tat" outperformed strategic competitors. In this game, the first player cooperates and then reciprocates the moves of the other player. Defection is met by retaliation, but cooperation is also reciprocated. This winning strategy "promote[s] mutual interest rather than … exploiting the other's weakness" (p. 130). Preconditions for the success of this strategy include the probability that players will recognize each other and again interact, conditions likely met in man's early evolutionary environment when kin lived in close proximity. Cooperation can also lead to in-group solidarity and ethnocentrism (Dawes, Orbell, Simmons, & Van De Kragt, 1986; Hammond & Axelrod, 2006). Results like these are bolstered by studies in evolutionary psychology which posit evolved mental biases for factors such as kin recognition (Antfolk, Lindqvist, Albrecht, & Santtila, 2014) or cheat detection (e.g., Barkow et al., 1992) as enabling mechanisms.

Peter Corning argues that an understanding of the synergistic properties of complex systems will contribute to our understanding of political behavior and outlines the normative consequences of such an understanding. In some ways, Corning's journey has taken him back to the place where he started. In 1977, Corning wrote that human nature revolves around meeting basic survival and certain higher level needs. Politics encompasses both the individual and social efforts to achieve these ends. Since then, Corning has pursued nothing less than an understanding of how the natural world is constructed and how this construction has produced the complex natural and social arrangements we experience. Thus, in the *Synergism Hypothesis* (1983), he traces the emergence of complex systems to the combinatorial effects of more elemental entities. When combined, these elements produce a whole that is more than the sum of the parts, much as table salt is produced by combining light metal sodium with chlorine gas, elements which alone are harmful for human consumption but when combined produce a socially useful product.

In evolutionary theory, functional synergy refers to the "combinatorial or cooperative effects that have had positive consequences in relation to various aspects of the problem of survival and reproduction" (Corning, 1983, pp. 4–5). The unit of analysis is relationships and the purpose is discovery rather than the postulation of covering laws or mechanical properties. A holistic Darwinism, as opposed to neo-Darwinian theory, is not limited to understanding selection at the level of individual units like the genes but sees selection as an evolutionary force involving both competition and cooperation.

Corning identifies examples of naturally occurring symbiosis, mutualism, and cooperation from the cellular level to the phenotypic level across a range of species from bats to humans. This does not mean that competition for resources is not part of the evolutionary story, but it is not the whole story. Often cooperation enhances the ability of in-groups to compete with outsiders. Corning's holistic account also holds that causation is a two-way street. Not only do the genes influence behavior but "functional (synergistic) properties of the whole become a selective screen – a significant influence on the differential survival/reproduction of the parts" (2008, p. 32). Accordingly, group behavior can exert downward pressures that alter the odds of survival for group members, their genes, and their progeny. Because humans (as well as many other species) live in groups, the nature of their relationships impacts survival and reproduction. For example, cooperation in certain circumstances enhances the fitness of conspecifics. Some of these cooperative relations are so successful in the joint survival enterprise of their practitioners that they engender policing strategies for enforcement. "[An] implication of this insight is that political theory can now be connected directly to evolutionary theory; insofar as politics and government are related to the problem of survival and reproduction in any species, there is a common functional basis" (Corning, 2008, p. 34). Human political systems may thus be viewed as a means for achieving collective survival.

In *The Fair Society* (2011), Corning builds on these themes and suggests a blueprint for a biosocial contract based on our evolutionary heritage of reciprocity, mutual aid, and targeted altruism. Returning to the notion of basic human needs, he suggests 14 primary needs that societies should secure for each of its members. These are not only tied to survival like the need for water, sleep, and procreation, but also to higher level needs like providing a supportive environment for the maintenance of social relations and even the promotion of mental health. He proceeds to suggest three principles of fairness in meeting these needs for the collective survival of the group (2011a, p. 154):

- Goods and services must be distributed to each according to our basic needs
- Surpluses beyond the provisioning of our basic needs must be distributed according to merit
- In return, each of us is obligated to contribute proportionately to the collective survival enterprise in accordance with our ability

Current ideological monoliths of capitalism and socialism are shown to be inadequate to the task.

Impact of First-Generation Biopolitical Theory on Mainstream Political Science

Assessing the intellectual impact of three decades of biopolitics on the mainstream of the discipline, Somit and Peterson wrote that only one verdict is possible: "biopolitics must be viewed as basically unsuccessful" (2000, p. 569). The same can be concluded about the first generation of biopolitical theory. While the work of first-generation biopolitical theorists occasionally found its way into mainstream journals and books, it remained the preserve of very few practitioners outside of the biopolitical community. There are several reasons why this might be so.

First, despite attempts by biopolitical theorists to put concerns about the "naturalistic fallacy" to rest, mainstream theorists appear not to be convinced. The position of Holly Lawford-Smith is illustrative:

> Most people accept that (some) facts place constraints upon normative claims. "Ought" implies "can," so by contraposition "not-can" implies "not-ought." But most people

> deny that (any) facts entail normative claims. One cannot derive an "ought" from an "is." (2012, p. 314)

This is especially true for universal claims, but it applies no less to contingent claims like those that posit "if we want to survive, flourish, etc. we should …"

> "If I want to quench my thirst, then I should drink some water." I want to quench my thirst. I should drink some water. But that is a prudential imperative, it comes from the fact that I want to quench my thirst. No one thinks that I ought morally to drink some water. (2012, p. 317)

Second, even if we limit discussions of moral imperatives to the negative constraints imposed on our behaviors by biological fact, the nature of those "facts" is often in question. This argument is a refrain of the one posed by Richard Lewontin (1979) against sociobiology in the 1970s when he characterized evolutionary explanations of behavior as "just so stories" that could not be verified. We have come a long way from sociobiological "just so" stories, but skepticism remains. Given the often divergent perspectives of life science subfields, facts and methods are often contested. Debate continues, for example, over the importance of kin selection versus group selection, the latter of which posits that, under some circumstances, the characteristics and behaviors of groups are implicated in fitness enhancement. This debate was recently reopened by E. O. Wilson who had previously championed kin selection explanations (see, e.g., Wilson & Wilson, 2007; and for an opposing view, see Dawkins, 2012). Similarly, debate continues over whether "downward causation" can be explained away by a proper understanding of functional reductionism, rendering the former concept irrelevant and permitting complete explanation on the basis of lower-level causal processes alone (see, e.g., Campbell & Bickhard, 2011; Kim, 2006; Marras, 2006). These debates are important for biopolitical theory since these ideas are incorporated into some of the theories we have reviewed.

In his critique of biopolitical theory, Mika LaVaque-Manty takes note of continuing debate over biological "fact" and concludes: "These differences suggest that perhaps political theorists and others interested in understanding what restrictive constraints nature imposes on human political possibilities ought to wait on the sidelines while scientists sort things out among themselves and decide which of these approaches ultimately gets vindicated" (2006, p. 9). In addition, since biopolitical theorists embrace a multilevel, multicausal approach to understanding behavior, it is difficult to know which facts are dispositive in a particular case and consequently which moral route to take. For example, anorexia may be hardwired in some individuals but females in a gendered society are often driven to this state by social cues. Given these competing (and perhaps overlapping and even reinforcing) "facts," and assuming that this is an unhealthy behavior, it is difficult to know whether we "ought" to treat this as a health problem, a social problem, or both. Which route we take depends on which "facts" we believe supervene.

Of course, failure to convert does not mean a failure of enterprise. In their assessment of the impact of biopolitics 30 years out, Somit and Peterson presciently note that "scientific paradigms are often changed less by intellectual conversion than by generational succession. In the long run, then, as political science moves to keep pace with its sister disciplines, biopolitics will almost certainly achieve its basic intellectual objective" (2000, p. 570). A second generation of biopolitics practitioners has now emerged in the field, and, despite their relative independence from first-generation scholars, they should, nevertheless, help fulfill the goals of the subfield's founding generation.

SECOND-GENERATION BIOPOLITICS

If first-generation biopolitics went largely ignored by the mainstream, second-generation biopolitics has, by comparison, taken the mainstream by storm. The breakthrough seems to have come in 2005 when John Alford, Carolyn Funk, and John Hibbing published "Are Political Orientations Genetically Transmitted?" in the *American Political Science Review*. The piece spurred numerous follow-up articles by members of this team as well as their protégés and colleagues, numerous accounts in the press, and spirited debate in political science journals ranging from praise (Sigelman, 2006) to skepticism (Charney & English, 2012, 2013), to questions of relevance for the discipline (Schlozman, 2013; see also Bartels, 2013).

Genopolitics

> Genopolitics rests on the foundation of twin study methodology .... The underlying foundation of twin studies is rooted in biometric theory and uses structural equation modeling of the covariance between monozygotic and dizygotic twins to evaluate the contribution of latent genetic and environmental influences on the trait of interest (Medland & Hatemi, 2009). Estimates for the parameters of these latent factors are

> then assessed for goodness of fit, often using maximum-likelihood criteria, though other approaches can be employed. Extended family designs build on the basic twin study but incorporate knowledge about the genetic relationships with other family members to provide additional leverage in analysis. (Fowler, Loewen, Settle, & Dawes, 2011, p. 215)

In their breakthrough article, Alford et al. (2005) drew upon data from an Australian twin study derived from surveys of several twin cohorts (both monozygotic and dizygotic) and their first-degree relatives and spouses (*N*>20,000) and a similarly drawn American sample of over 24,000. Using this massive data set, the authors found that genetics plays a significant role in explaining political attitudes and ideology. Their analysis used items in the Wilson–Patterson (W–P) Attitude inventory, which includes questions related to conservatism. The researchers sifted the data to determine the impact of heredity, shared environment (parental socialization), and nonshared environment (unique differences in diet, disease exposure, peer influence, etc.) on political attitudes while controlling for sex of twin pairs in order to exclude male–female differences that might deflate correlations for DZ pairs relative to MZ pairs. The authors found that correlations for the 28 attitude items surveyed were higher for MZ pairs than for DZ pairs and in all cases the differences were significant at the .01 level (2005, p. 158). The impact of shared environment exceeded that of heredity for only four items and the mean estimate of heritability for all 28 items was .32 compared to a mean estimate of shared environmental influence of .16 (at p. 159). An estimate of the shared environment for all 28 items was .53 or roughly half of the overall variation. Summarizing the results, the authors state: "… shared influences (genetic and environmental) account for about half the variation in these political reactions, with unique individual and environmental factors accounting for the remainder. Within the half that is accounted for by shared influences, genetic influences, in contradiction to behavioralist expectations, are roughly twice as influential as environmental influences" (at p. 159).

Collapsing all 28 inventory items into a single index and computing correlations against both sets of twins, the estimate for heritability is .43, higher than for any individual item; the estimate for shared environment is .22; and the estimate of unshared environment falls to .35. The decline in the role of the unique individual environment seems sensible, the authors conclude, because the index cancels out idiosyncratic features of individual survey items (at p. 160). A separate examination of attitudes toward political party found that hereditability is rather low (.14) while shared environment accounts for a much greater share (.41). Yet, affect toward party (measured by pro–con reactions) results in a heritability estimate of .31 and a shared environment estimate of .17, similar to results for the overall composite index. The authors reason that while the party one chooses may depend more on environment than genetics, the strength of attachment to a party may depend substantially on genetic factors.

The authors go further. Because we cannot dismiss the possibility that mates of dizygotic twins (who share only half of their genes) share genes for traits of interest, the impact of assortative mating must be factored in the equation. "Any measurable tendency toward assortative mating on political orientation will push up the DZ twin correlation while leaving the MZ correlation unaffected, and this reduction in the MZ-DZ gap will have the related effect of lowering estimates of heritability" (at p. 161). Using a subsample of the American data set that includes parents' individual responses to the same W–P items used with the twins, the authors conclude: "When the impact of parental agreement is partialed out, the average estimate of heritability rises to .53 and the average estimate of the impact of shared environment drops to .11" (at p. 162). Results were compared with the data set from Australia and the authors found the picture to be "remarkably similar" (at p. 163).

Alford, Funk, and Hibbing conclude that with the exception of party identification, political attitudes "are influenced much more heavily by genetics than by parental socialization" (at p. 164). The results, they argue, illustrate that there is a difference between "inherited attitudes" and acquired attitudes. The former, something like broad political orientations to social issues dealing with authority, morality, gender, and race, will be more resistant to change and find expression as "gut reactions" to political inquiries. They will also be more difficult to manipulate. The findings are of significance for those studying ethnic and social cleavages. Certain phenotypes, the authors report, are characterized by strong suspicions of out-groups (e.g., immigrants); a yearning for in-group unity and strong leadership, especially if there is an out-group threat; a desire for clear, unbending moral and behavioral codes; a preference for swift punishment; a willingness to tolerate inequality; and a pessimistic view of human nature. The other phenotype is relatively more tolerant of out-groups, more likely to take a context-dependent approach to human behavior, inherently optimistic about human nature, suspicious of hierarchy, has an aversion to inequality; and takes a more rehabilitative view toward punishment.

As mentioned, the Alford, Funk, Hibbing (AFH) article set off a firestorm of debate. However, it also spurred the publication of a rash of related twin studies of political attitudes and behaviors (for an excellent review of these, see Hannagan, 2011). In response to methodological critics (e.g., Charney, 2008a, 2008b), some of the follow-up studies employed more sophisticated modeling under a maximum likelihood framework (MLF) to infer the relative importance of the correlations between observed traits of MZ and DZ twins relative to underlying genetic and environmental impacts. Reexamining the AFH data using MLF, Hannagan and Hatemi (2008) found that while there are some sex differences in the magnitude of the variance components, "the more sophisticated methodology and more parsimonious explanation suggest that genes play an even stronger role in political attitudes [drawn from the W-P Attitude Inventory]. In other words, AFH provided a more conservative estimation using the simpler method" (2008, p. 330). Hatemi et al. (2009) confirm AFH findings with regard to party identification, noting that twin data confirm that the origin of party identification is almost exclusively cultural. However, "[P]artisan intensity is heavily influenced by genetic liability but influenced very little by familial socialization" (Hatemi et al., 2009, p. 598).

Attitudinal studies (Eaves & Hatemi, 2008) found that MZ twins have more highly correlated beliefs about abortion and homosexuality than DZ twins. Genetic effects accounted for 50 to 70 percent of variance, with lesser amounts for environmental experiences and social learning. Religious attitudes were found to be largely determined by environmental factors (Eaves, Hatemi, Prom-Womley, & Murrelle, 2008), but the authors speculate that religious training, especially in one's formative years, may play an important fitness enhancing role by keeping young adults from engaging in risky behaviors.

Hatemi, Medland, Morley, Heath, and Martin (2007) turned their attention to political behavior, specifically the impact of genetics on vote choice. Using the Australian twin study data and self-reports of voting behavior, the team found only a modest genetic influence on vote choice. Still, they argue that vote choice could be explained by shared genetic influences related to political attitudes. They also reiterated that rational choice models used by mainstream political science present an incomplete understanding of political behavior because they fail to acknowledge genetic differences among individual actors.

Examining genetic influences on voter turnout, Fowler, Baker, and Dawes (2008) found that 53 percent of the variance in turnout behavior could be attributable to genes. The team matched individuals drawn from the Southern California twin registry against publically available voter history records in Los Angeles County. They corroborated their results by using self-reported political participation rates from a national sample drawn from the National Longitudinal Study of Adolescent Health (Add Health). They conclude: "In summary, both Add Health replications yield estimates of heritability that are similar in magnitude to the 53% estimate for heritability in the Los Angeles sample, suggesting the heritability of political participation is robust" (2008, p. 242). The authors suggest that the next step in genopolitics is to identify specific genes as candidates for study as contributors to political behavior.

Following up on this promise, Fowler and Dawes (2008) narrowed their search for genetic markers of voting behavior by focusing on two genes related to neurochemical transcription, MAOA and 5HTT. Since it is known that these genes impact prosocial behavior, the authors hypothesized that they might help explain differences in voter turnout. Polymorphisms of each gene exist in high/low and long/short alleles, respectively. A high MAOA polymorphism and long 5HTT polymorphism better regulate stress and individuals possessing these were hypothesized to be more likely to vote. Using data from the ADD Health twin studies including self-reports of partisan ID and voting behavior in the 2000 election, the authors found that only the high version of MAOA was associated with voter turnout. Those with this form were substantially more likely to vote than those with the low form. Individuals with long 5HTT were also more likely to vote but only if they also participated in religious activities. In summary:

> Even after controlling for factors known to influence turnout, having a high MAOA allele raises the likelihood of voting by about 5%. Among people active in their religious organizations, having a long 5HTT allele raises the likelihood of voting by about 10%. We theorize that since low efficiency MAOA and 5HTT alleles limit the degree to which individuals are socially oriented, these alleles inhibit their desire or ability to participate in the political process. (Fowler & Dawes, 2008, p. 588)

While these findings provide a plausible genetic link to political behavior, they also point clearly to the importance of gene–environment interaction.

In 2013, Deppe et al. replicated the 2008 Fowler and Dawes study in response to methodological concerns by Charney and English (2012). Using methods designed to surmount these concerns, they confirmed the Fowler and Dawes findings when using self-reported voting as the dependent variable. But, consistent with criticisms by Charney and English, the relationship between the candidate genes and turnout was shown to be fragile. In fact, substituting a more reliable and robust measure of voting frequency for self-reported political participation and using state-of-the-art categorization of the genotype, the relationship disappears entirely (Deppe, Stoltenberg, Smith, & Hibbing, 2013, p. 380).

In 2009, Dawes and Fowler studied the impact of the A2 allele of the D2 dopamine receptor gene on partisanship and voting. The authors propose that the empirical link between the D2 dopamine receptor and the DRD2 gene, along with the known functions of dopamine in the brain, suggest at least two channels through which the A2 allele might influence partisanship. "We hypothesize that improved ability to form social attachments and/or improved cognitive function, both of which have been shown to be associated with the A2 allele, increases the likelihood an individual with the A2 allele will form and/or maintain an attachment with a political party" (2009, p. 1164). Utilizing ADD Health twin data once again and employing two different test designs, one controlling for the impact of a number of social and demographic factors and the second using a family-based test of association comparing allele frequency among siblings displaying or not displaying the trait, the authors found a significant association between the A2 allele and partisanship. There was no direct effect between the DRD2 gene and voting, but there was an indirect effect mediated through partisanship. The authors argue that genes play a role in partisan attachment and help explain why partisan attachment runs in families. In turn, partisan strength helps explain voter turnout. Because the D2 dopamine receptors have been implicated in social anxiety and stress, they suggest the road from genes to partisan behavior may run through human emotions (see Chapter 5).

Settle, Dawes, and Fowler (2009) reported similar results using a sample of 2,000 pairs of twins attending a festival in Ohio. The authors inquired into partisan attachment and strength of partisanship using a scale similar to that employed in the American National Election Studies (ANES) studies. The results, they report, suggest that heritability generates about 46 percent of the variance in strength of partisan attachment. Unshared environment plays a lesser role and shared environment appeared to have no appreciable impact (2009, p. 605).

> While arguing that partisanship can no longer be seen as purely a cultural phenomenon, they do point out that "our study opens the door for an examination of factors that could serve as mediators for the gene-attachment mechanism. It seems likely that we are tapping into a general predisposition toward attachment, interest, or engagement and that environmental factors can play an important role in channeling that predisposition into behavior." (2009, p. 609)

Previous research by Settle et al., for example, found that "the number of friendships a person has in adolescence is significantly associated with liberalism among those with DRD4-7R. Among those without the gene variant, there was no association. In other words, it was the interaction of a particular gene and a particular environment that mattered" (2009, p. 609).

THE NEXT FRONTIER

Genopolitics has opened some exciting frontiers in biopolitics and it has awakened the mainstream of our discipline to a discussion of the biological bases of behavior that the first generation of biopolitics advocates were not able to do. Yet, genopolitics is still in its infancy and largely uncharted. There remain methodological as well as ideological critics whose concerns have not been totally satisfied. There also appears to be some inconsistency among findings. But even if political scientists are inclined to pay attention to the findings of genopolitics and even if methodological issues can be resolved, there remain significant questions about exactly what the myriad genetic findings really mean.

While genopolitics practitioners in large part have done their best to add the relevant caveats that single genes do not simply translate into political behaviors and to caution that we are far from certain how genes coding for proteins impact the function of neurons and synapses in ways that cause the organism to behave in a certain manner, such assurances are not likely to convert members of a discipline who have spent their entire lives learning about how *phenotypes interact*. Nor should they be. Ultimately, genopolitics arises from personality studies. This literature has a long and venerable history and has informed our own discipline at times (Adorno, Frenkel-Brunswik, Levinson, & Sanford, 1950; Costantini & Craik, 1980; Greenstein, 1975; Jennings & Markus, 1984; Jennings & Niemi, 1981; Wilson, 1973). But advocates of genopolitics have often made a leap not simply from genes to behavior but from personality traits (for which there may be a legitimate genetic basis) to specific forms of political behavior with little evidence connecting the separate domains. For instance, do we really know that voting is a form of prosocial behavior (Fowler & Dawes, 2008)? Much voting – especially in referenda for social causes – is a statement opposing social conventions or institutions. Moreover, does positing the heritability of prosocial behavior really defeat Downsian explanations based on costs and benefits? Or, does a genetic predisposition toward openness to novelty really translate into support for abortion (Eaves & Hatemi, 2008)? Does social anxiety linked to alleles in one's genes really mean one is less likely to vote or write one's Congressional representative on an important issue (Dawes & Fowler, 2009)? Missing from this literature is an account of how the personality traits associated with the genes translate into specific political behaviors.

Mondak, Hibbing, Canache, Seligson, and Anderson (2010) speak directly to these concerns and point to the road forward. Taking seriously the concern that it is necessary to establish clear connections between personality traits and political behavior, they utilize as a touchstone what they call the Big Five traits (see Chapter 4) from a well-recognized model in psychology (Goldberg, 1990; McCrae & Costa, 1987, 2008). The model provides a comprehensive, hierarchical picture of trait structure with the following dimensions: openness to experience, conscientiousness, extraversion, agreeableness, and emotional stability. It has seen some initial use in political science (e.g., Gerber, Huber, Raso, & Ha, 2009; Mondak & Halprein, 2008) and the authors believe it provides a parsimonious representation of the bulk of trait structures applicable to political science research. Substantial research in psychology exists establishing the hereditability of the Big Five traits (see, e.g., McCrae, Jang, Livesley, Riemann, & Angleitner, 2001; Riemann, Angleitner, & Strelau, 1997). In these studies, hereditability estimates range from .46 to .56 but rise to .66 and .79 when corrected for measurement error (Mondak et al., 2010, p. 89). Similarly, Boreknau (1992) found virtually no impact of the unshared environment on the Big Five structure. The lesson we derive from these studies, Mondak et al. conclude, "is that biological forces account for most of the variance in personality traits" (2010, p. 89).

Once hereditability of the Big Five traits was established, Mondak et al. (2010) attempted to tie specific forms of political behavior to these traits. To do so, they conducted a series of telephone surveys shortly before and after the 2006 mid-term elections. Respondents were asked to rate themselves on 10 personality items using pairs of words like "tense" versus "relaxed." Scales for each of the Big Five traits were then constructed from these accounts and individual scores assigned. Respondents also were asked to report on their political activities ranging from attending public meetings, to contacting a member of Congress, to trying to convince others to work for a campaign, and on the size of their social networks. The impact of contingent factors such as measures of civic knowledge and efficacy were also taken into account. In the end, four of the Big Five traits were found to be associated in some way with political activity. The major findings of this study demonstrate that the:

> ... impact of extraversion on participation is contingent on the nature of the participatory act in question, that the effects of openness are mediated partly by political information, that conscientiousness operates in concert with other attitudes regarding civic duty, and that personality moderates the effects of network size on a person's exposure to cross-cutting social communication. (Mondak et al., 2010, p. 101)

The authors reported inconsistent results regarding emotional stability both within and across studies and that agreeableness interacts with other personality factors like extraversion and contextual factors like the size of one's social network.

While the authors believe that their findings strongly support the impact of biology on personality and believe that biology helps explain how personality traits contribute to political behavior, they present a far more nuanced and complex picture of political behavior than genetic studies alone provide (see Fig. 2). They are also realistic about the road ahead, arguing that more effort is needed to tease out the impact of situational and genetically mediated dispositional factors contributing to political acts. "Numerous

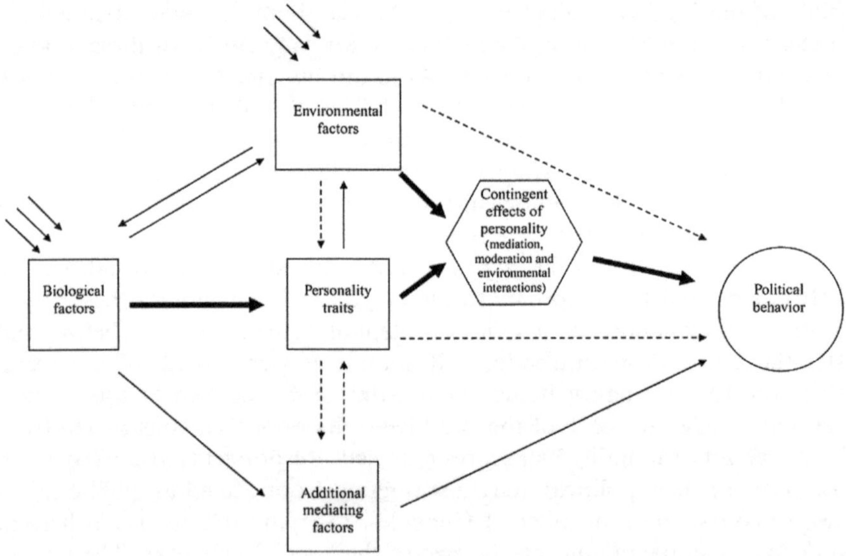

Fig. 2. Personality in Context: Personality Influence on Political Behavior. *Source*: Mondak et al. (2010, p. 88).

conceivable intervening processes must be assessed if the mechanisms underlying personality effects are to be identified" (Mondak et al., 2010, p. 104). While they admit that they have operationalized "environment" in a very broad manner, future work "must endeavor to differentiate more systematically between mediating and moderating effects, and must define and categorize nonpersonality factors more precisely" (at 103).

THE FUTURE OF BIOPOLITICAL THEORY: BRINGING THE GENERATIONS TOGETHER

While the pioneering generation of biopolitics scholars made great strides in (re)introducing the vocabulary of biology into the discipline of political science, they were less successful in attracting wide interest among practitioners in the discipline. Yet, first-generation biopolitics scholars did make progress in broadening the scope of disciplinary discourse and positioning the study of politics within the broader intellectual framework of evolutionary theory. As a result, politics could no longer be seen as a novelty of the human species, but instead as an adaptation drawing its foundation from the rudiments of coalition formation, hierarchy, and resource distribution common to other species as well. Within this broadened scope, it became possible to ask normative questions related to the purposes of political life, the evolution of altruism, the origin of justice and fairness, and the duties to future generations.

While the second generation of biopolitics scholars made great strides in advancing empirical links between genetics and politically relevant personality attributes, attracting wide attention in the process, they have been less successful in tying their findings to broader issues in evolutionary and normative theory. They have filled in many of the details about the genetic contributions to political behavior that was hypothesized, but not delivered, by first-generation scholars. But they have done so largely by ignoring the contributions of their forebears. Although a few genopolitics articles have made passing mention of first-generation biopolitics, there has been virtually no attempt to tie their present endeavors to earlier work, thus creating a generational divide that has been denounced by Gary Johnson (2011, p. 52). Also, by not doing so they run the risk of simply adding one more set of variables to regression analyses without an appreciation of what these variables say about the role of politics in the life of the species.

When asked by a *New York Times* columnist about the impact of recent findings in genopolitics, Gary Jacobsen commented, "The evidence that there is a genetic influence on some [political] attitudes and behaviors seems quite strong." But he added: "What we do with this knowledge is another matter. How do we look at public opinion differently knowing that some of what we measure has a genetic basis? I am not sure what the answer is" (cited in Edsall, 2013). Kay Lehman Scholzman reacts similarly in a symposium in *Perspectives on Politics*. She says political scientists should be concerned about two questions: "Why?" and "What difference does it make?" While genopolitics is responsive to the first, it has not substantially answered the second. "As a political scientist, I need a fuller explanation as to what difference it makes for public life that being inclined to cringe from the slimy and germy [one measure used by Hatemi to test for genetic predisposition to certain attitudes] is associated with ideological conservatism … [T]he field of biology and politics is vulnerable to this charge" (2013, p. 491). Unmoored from its evolutionary foundations, genopolitics risks becoming irrelevant.

In many ways, the work of each generation of scholars is incomplete. The first generation outlined the scope of politics within an evolutionary paradigm, but did not supply the tools to fill in the details of how nature and nurture interact. The second generation has added myriad details, but has failed to locate them within the broad life science framework within which it belongs. A glance at Figs. 1 (Masters) and 2 (Mondak et al.) presented earlier in this chapter, however, reveals many potential points of contact between these generations of researchers and may point the way to the next era of biopolitical research. Fig. 3 provides a tentative sketch of the ways in which these figures intersect.

The Masters' framework is broad but lacks detail about proximate mechanisms; the Mondak sketch is detailed yet incomplete. It lacks consideration of ultimate causes and the feedback loops that connect both levels of causation in an evolutionary context (see Losco, 1981). The Mondak chart begins with genetic information retrieved from the gene pool and ends with individual behavior. It does not, nor does it purport to, indicate what evolutionary factors favored the particular constellation of genes that gave rise to specified behaviors in particular environments. Nor does Mondak indicate the fitness consequences of the behaviors in question for the individual, his/her kin, or for future generations. By incorporating feedback loops that trace the impact of certain behaviors to social systems, the gene pool, and the entire human species, Masters forces us to consider the reasons genes coding for specific personalities

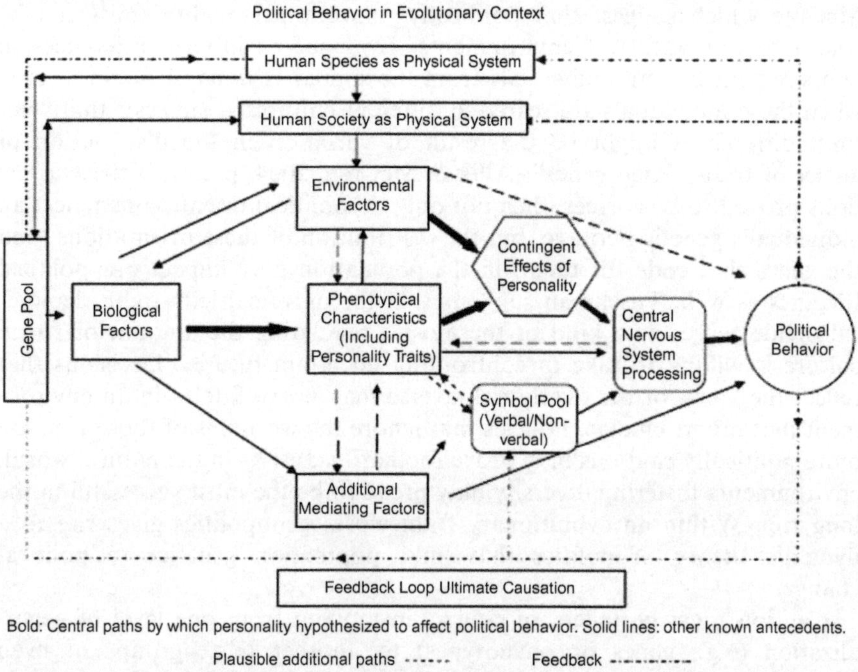

Fig. 3. Political Behavior in Evolutionary Context.

and behaviors have been selected in the first place, their distribution throughout populations, and their continued impact on fitness for the individuals and groups displaying them. A unified evolutionary theory of politics then, like the one sketched below, will necessarily relate genetically mediated political behaviors to the evolutionary forces which gave rise to their selection and to their continued impact on fitness.

Evolutionary anthropologist Avi Tuschman applies a perspective somewhat like the one outlined in this figure. In his recent study (2013), he hypothesizes that the extent to which a particular society embraces change (left-leaning ideology) or clings to tradition (right-leaning ideology) depends in part on the distribution of genes coding for politically relevant personality traits like extroversion or introversion within the population. This is similar to the claims made by genopolitics adherents like Alford et al. (2005) and Mondak et al. (2010). But Tuschman goes further by positing that these characteristics have deep roots that extend back generations in our evolutionary history. He cites studies like those by Allick and

McCrae which suggest that politically relevant personality traits can be traced to our ancestral environments. Evidence of inborn differences in temperament among individuals from the similar regions of origin – even when these individuals are reared in different cultures – suggests that "cultural differences might be the result of variation in the distribution of alleles of trait-related genes" (Allik & McCrae, 2004, p. 24). If these assertions proved to be correct, then not only do political orientations reflect an individual's genetic heritage, but the distribution of these orientations (and the genes that code for them) in the population may impact our political fortunes as well. Tuschman suggests that the universal left–right ideological divide serves as a kind of thermostat regulating the amount of risk a culture is willing to take in confronting uncertain futures. Decisions that reflect the views of the overly risk-averse may prove fateful in an environment that favors change. Policies that ignore the warnings of those who are more politically cautious may prove foolhardy. Just as in the natural world, environments fostering diversity may prove to be the most successful in the long run. Within an evolutionary framework, genopolitics gives rise to a dynamic theory of politics that links population genetics to political change.

Caution must be taken, of course, in moving from one level of generalization (e.g., genes or phenotypes) to another (e.g., groups or even whole societies). Moreover, much of the evidence emerging from genetic studies remains contentious. Nevertheless, the merger of two generations of biopolitical scholarship may prove fruitful in the long run. Both genopolitics and evolutionary theory are necessary to fully understand political behavior and should be integrated in future research. Genetically mediated attitudes may reveal evolved patterns of thought that serve(d) useful fitness functions. Here, genopolitics could benefit from incorporation of findings from evolutionary psychology. How do hereditable attitudes reflect the thought biases that evolutionary theorists believe to reflect our longer-term evolutionary heritage? In turn, a fully developed genopolitics of behavior can provide keys to the proximate mechanisms by which ultimate evolutionary functions are impacted. Do genetically mediated behaviors and attitudes enhance fitness? If so, why do alternative forms persist? Do these, too, have some fitness consequences for the individual or the group? Such a synthesis would be useful in the arena of policy studies as well. If civic engagement were to be found to enhance fitness, for example, then we may wish to discover strategies that work best for individuals with different genetically mediated personality traits.

CONCLUSION

Aristotle called man the "*zoon politikon.*" By this, he did not mean simply that we are social animals. After all, *homo sapiens* is simply one of a number of social species about which Aristotle was aware. What makes us distinctive and truly *zoon politikon* is that we are able to make moral judgments by virtue of our power of speech and thus to regulate our behavior.

> Now that man is more of a political animal than bees or any other gregarious animals is evident. Nature as we often say makes nothing in vain, and man is the only animal whom she has endowed with the gift of speech. And whereas mere voice is but an indication of pleasure or pain, and is therefore found in other animals ... the power of speech is intended to set forth the expedient and the inexpedient, and therefore likewise the just and the unjust. (Politics I (ii))

Moral behavior takes place within a social context, a public life, in which behavior is regulated by moral thinking about our relationship with others in the community. If, as the proponents of genopolitics assert, many of the attitudes we hold about what is fair, how we treat kin as well as strangers, our toleration for difference, etc., are heritable, then the key to understanding politics may lie in our evolutionary past. Our inherited predispositions coupled with our distinctive capacity for language and judgment function to mediate and regulate social life in ways that are analogous to hierarchies, threat displays, and bonding patterns in related species. When placed in an evolutionary context, genopolitics gives us a deeper appreciation for the extent to which our behaviors, although distinctive as a species, have been molded by our biological past and how they may impact future generations. At the same time, a biologically oriented political science can connect our discipline more closely to its intellectual roots.

CHAPTER 3

INTRODUCTION TO METHODOLOGICAL ISSUES IN BIOPOLITICS

Patrick Stewart

Those studying the influence of biology on social and political behavior find themselves in the final stage of Haldane's "process of acceptance," where knowledge goes from (stage 1) being worthless nonsense, to (stage 2) an interesting but perverse point of view, through (stage 3) being true but unimportant, to the final stage (4), where critics always said so. While there has been movement away from the early stages, which were much less benign than characterized in Haldane's formulation (Somit & Peterson, 1999), biological approaches are only just now moving into general acceptance by the field of political science and its subdisciplines (Hatemi, McDermott, Eaves, Kendler, & Neale, 2013, p. 6). This gradual acceptance, with interdisciplinary efforts exploring political and social issues, are beginning to bear fruit in the top political science journals, although other disciplines seem to have embraced biological approaches to politics earlier and more fully (Hatemi & McDermott, 2011), and has started to resonate with the general public. This acceptance may be seen in the recent burst of popular press books that approach political and social issues from a biologically aware and astute perspective (Hibbing, Smith, & Alford, 2013; Saad, 2011a; Tuschman, 2013; Van Vugt & Ahuja, 2011).

Politics and the Life Sciences: The State of the Discipline
Research in Biopolitics, Volume 12, 67–99

ISSN: 2042-9940/doi:10.1108/S2042-994020140000012002

However, due to institutional inertia, this knowledge is being incorporated into the disciplines that often have provided the fiercest critics of biological approaches. These criticisms often appear to be theoretically focused by decrying evolutionary theory as Social Darwinism or just inappropriate for addressing social questions; however, the methods used in biopolitical inquiry may prove to be even more of a challenge to institutional and disciplinary inertia. While methods and theory may be seen as intrinsically wedded, with advances in one leading to greater and more precise insights in the other, new methods systematically challenge belief systems in ways that theoretical advances tend not to, as data-driven findings are harder to dismiss than thoughtfully crafted insights. More specifically, the techniques used to gather biologically relevant information, and the research designs used to test inferences, are exceedingly advanced in comparison with the typical approaches used in social science research. Here, the traditional reliance on self-reports through public opinion surveys and analysis of secondary data is prone to both systematic and random error that occurs with this type of data. And while this data may be analyzed using highly complex statistical methodology, the approaches used can often obscure more than is revealed, due to relying on correlational case-study designs to draw inferences. However, this methodological hegemony is slowly being overturned by a broad array of data gathering methods and research designs that are informed by biological approaches and evolutionary theory. As is seen throughout this chapter, the biopolitical "revolution" is now fully flowering thirty years after the seed of inquiry was planted.

Regardless of the findings and insights provided by research, critics have and likely will remain wedded to their viewpoints for the foreseeable future. As pointed out by Thomas Kuhn in his seminal work *The Structure of Scientific Revolutions*, often disciplines will only change with time and replacement of one theoretical viewpoint by the next generation of scholars who embrace a new paradigm supported by overwhelming evidence (Kuhn, 1996). More to the point, while extremist viewpoints about human behavior are still in evidence, they are crumbling under the weight of evidence and the opening of cooperative connections across disciplinary boundaries. Examples of resistance to change proffered by the interdisciplinary nature of biological understandings of human behavior may be seen in two extremes of understanding humans and their actions. On the one hand, there are the constructivists who believe in the infinite malleability of humans by society and hence the inability to draw any conclusions about human behavior. On the other hand, there are those who posit the unfailingly predictable human, as seen in the rational actor models of economics and the disciplines that follow their intellectual lead.

Constructionist approaches to knowledge, at their extreme, posit that context determines all social behavior. Here, the major proponents of this approach engage in a critical perspective which denies the validity of knowledge (Liesen & Walsh, 2013; Stewart, 2013). While critiques of the fundamental tenets of knowledge provide value (Fletcher, 2013; Liesen & Walsh, 2013), critiques that deny meaning outside of context lack value as contexts are ever varying. A further complication with postmodern applications on scientific advances is that their critiques tend to be built on self-referential literature that is oftentimes very difficult to understand (Stewart, 2013). The reliance on social constructions, and with it literature solely reflecting this bias, suggests that postmodernists see society as influencing science, but apparently do not give much consideration to the production of knowledge through science or having a discernible impact on society.

Interestingly enough, recent interdisciplinary explorations of the effect of reading fiction suggests that the divide between the two cultures of science and literature, as presciently noted by C. P. Snow in his 1959 Rede Lecture, is being bridged (Snow, 1990). In two separate studies, scholars have found that both short- and long-form literature, and with it the construction of internal worlds, leads to changes in the connectivity of neural networks (Berns, Blaine, Prietula, & Pye, 2013), and enhanced "theory of mind" (TOM), in other words, the ability to read other people's affective states (Kidd & Castano, 2013). Furthermore, the production of social categories and explanations have a definite and powerful influence on political and policy choices (Ingram, Schneider, & DeLeon, 2007; Stone, 1997), albeit in a manner bounded by biological parameters (Harris & Fiske, 2006, 2007).

The other extreme may be seen as that of the rational actor model, where the actor is unchanging in response to context. As pointed out by Monica Capra and Paul Rubin, "economists had been heavily influenced by physics, which aims at finding unified theories for understanding the physical world" (2012, p. 322; see also Friedman, 1953). The advances made in the field of economics, and then applied to political science, tend to be based upon understanding deviations from the rational choice model of decision making at the individual level (Peterson & Lawson, 1989; Quattrone & Tversky, 1988), yet not necessarily better theories grounded in human behavior as it occurs. While this does not deny the contributions the rational choice theory has made, especially at the aggregate level, the persistent adherence to this theoretical model which continues to fail in accurately describing behavior – even when constrained to self-interested and competitive behaviors predominating in the business world – suggests a level of conservative dedication to the tenets of the "dismal science," especially when a more powerful, robust, and integrated option exists (Saad, 2007, 2011b). Furthermore, it suggests a blind dedication to assumptions about humans that may be considered harmful by prioritizing misanthropic behavior that does not take into account the social nature of humans.

The limitations of this model have long been appreciated by biologically oriented political scientists. As noted in Chapter 1, John Wahlke noted the sense of disappointment in "political behaviorism," disappointment expressed by critics and proponents as well, and characterized conventional work as primitive political behavioralism, urging greater use of biobehavioral science in order to overcome earlier limitations (1979). He quite accurately asserted that the distinctive feature of behavioral political science has been its adherence to the unified view of science (with physics as the model science), a position which sustains the quest for empirical theory. The major premise of political behavioralists "was a seemingly rigorous empiricism, often expressed as 'methodological individualism', i.e., insistence that the ultimate legitimate unit of observation in political study is the acting human individual, all other phenomena being regarded as 'merely' some collective expression of or inference from those observations" (Wahlke, 1977, p. 1). He pointed out that the major shortcomings of political behavioralism were the failure of micro-level studies "to articulate with appropriate 'macro-level' concerns of political science" and the overwhelming "reliance on an unrealistic and misleading model of the individual political actor, or, as earlier political scientists might have said, a defective conception of human nature." We concur with Wahlke's assessment. Indeed, it appears that political behavioralists are increasingly appreciating what biologically oriented scholars such as Wahlke proposed long ago.

BIOLOGICAL APPROACHES TO UNDERSTANDING HUMAN BEHAVIOR

What we are currently seeing in research using biological approaches is a broader appreciation for the interplay of the actor with their context, which also includes other actors, and with it the understanding that both the context and the actor influence each other in a recursive and interactive process. In other words, as humans we affect and alter our environments, but these self-same environments change us at the same time. As pointed out by Roger Masters, it is not simply nature versus nurture, nature and nurture provide interwoven threads in a whole cloth (Masters, 1989). The old nature (instinct, endogenous) versus nurture (learning, exogenous) debate between biologists and social scientists, respectively, is resolved by the evolutionary theoretical framework because it makes no sense to consider animal development and behavior in the absence of an environment. Likewise, the extreme of environmentalists' claims would posit an environment without any organism. In short, all phenotypes derive from the interaction of an organism's genetic potential with its environment, and behavior is as good a phenotype as any other (Barash, 1977). This point has been argued persuasively, in our view, on behalf of biopolitical inquiry by Losco (1981), Masters (1975), and White (1972). Recognizing the interactional nature of this phenomenon goes a long way toward reducing the tendency to see in sociobiology or biopolitics a new form of Social Darwinism. It also underscores for the social scientist the research task of integrating micro- and macro-level analysis.

Understanding politics often means organizing the world into understandable categories (Lowi, 1972). This organization may be so useful as to be relatively unchanging, or may be faulty to the point of needing the redefining of categories and their boundaries, and reorganizing knowledge.

Much of the success of and contentious consequences aimed at biopolitics is due to its transgressing of traditional academic boundaries and the guardians of these disciplines. This does not mean a chaotic view of science where anything goes, but rather an appreciation for the different levels and units of analysis that interact and how this interaction may inform understanding of political behavior. Here, biopolitics has taken the insights from the European School of Human Ethology, most notably Tinbergen's four questions (Tinbergen, 1963), and organized them into biobehavioral questions which in turn inform the biopolitical/biopolicy questions.

These four types of questions about behavior were posited by ethologist Niko Tinbergen in his seminal 1963 paper and consider proximate causation, ontogenetic development, phylogenetic roots, and ultimate causation (Tinbergen, 1963). The first two consider proximal causes of individual behavior. *Proximate causation* considers what motivates an individual to behave in a particular manner at a specific moment in time, asking questions concerning mechanisms that produce different types of behavior. *Ontogenetic development* looks at the roots of behavior; in other words, whether there are innate tendencies in upbringing or development of an individual over their lifespan that triggers particular behaviors. Both environmental and internal factors are considered in ontogeny, although both are intertwined and difficult to separate, especially as ontogeny occurs over an individual's lifespan (albeit with greater development occurring earlier in life).

The last two questions address ultimate causal factors and may be used to consider behavior at the individual level or at the population level with the understanding that data collected at one level of analysis may not provide a valid basis for generalization at another level (Peterson & Somit, 1982). The third question considers how behavior develops in a species over its history and analyzes its *phylogenetic roots* by considering differences in behavior among several species. *Ultimate causation* considers the functional cause of a behavior in terms of how it promotes (or alternately hinders) the reproduction and survival of a species, or group of organisms (Barrett, Dunbar, & Lycett, 2002; Lehner, 1996; Tinbergen, 1963), for instance, some behavior might benefit the group/species engaging in it in terms of obtaining and defending resources, which in turn affects rates of reproduction.

These four biobehavioral questions are driven by an appreciation for understanding the importance of evolutionary thought to observing and characterizing political behavior. Here, nonhuman mammalian, especially primate, species provide fundamental insight, and still present models of behavior for human political activities. This biobehavioral perspective not only draws heavily from classical ethology, it increasingly draws insights from such areas of inquiry as the brain sciences, behavioral ecology, evolutionary psychology, and genetics, among other fields. Furthermore, it differs from traditional perspectives of political scientists that focus solely on proximal influences on political behavior by considering four different types of questions. These questions are separate, yet not mutually exclusive and consider not just proximal causation by looking at the behavior of individuals, but also ultimate causation, which considers both individual- and population-level behaviors (Lehner, 1996).

POLITICS, POLITICAL SCIENCE, AND BIOLOGICAL EXPLANATIONS

This ultimately leads to the question of "what is the question" to be answered. Political science is undoubtedly about the pursuit and exercise of power. However, power is such an abstract and amorphous concept that it is rendered nearly worthless as an organizing construct. At the same time, the unspoken "who" that exercises this power leaves us unclear as to the unit of analysis. Specifically, the assumption often made is that politics is about current events, with the focus on the players, whether individuals, groups, or nations. As a result, the focus tends to be on the low-hanging and ripe fruit offered by the obvious contestants engaging in political action in terms of support for them through public opinion (Page & Shapiro, 2010), votes (Campbell, Converse, Miller, & Stokes, 1960), and financial support (Adkins & Dowdle, 2002; Dowdle, Limbocker, Yang, Sebold, & Stewart, 2013). The problem with this approach is that the focus is not only upon proximate explanations for behavior, but that methods used are comparatively myopic by adhering to disciplinary standards and ease of access to data sources (Sears, 1993). As a result, advances tend to be based upon improvements in statistical and analytical methodology as opposed to measurement, and are apt to be marginal gains and unlikely to challenge the predominant paradigm (Kuhn, 1996).

The clear starting point for political science ought to be that humans are eminently social animals. Like other social species, they rely on cooperation with conspecifics to survive, reproduce, and thrive in a range of environments. However, unlike other social species, cognitive capacity has given humans the ability to not only engage in larger social networks but also to hierarchically organize these networks into social structures. This ability is arguably what sets humans apart from other species. Therefore, as pointed out by Heinz Eulau:

> The political behavior of the individual person is the central and crucial empirical datum of the behavioral approaches to politics .... The political behaviorist concentrates on the behavior of individuals whose interactions and transactions make up collective behavior, even if he is concerned with describing and explaining the actions of groups, organizations, or other large collectivities. Groups, organizations, or nations have no independent status apart from the conduct of the individuals who are related by behavior towards each other in certain ways. This does not mean that groups, organizations, or other formations are not real and meaningful units with structural properties and functions of their own. They certainly are. In fact, the great bulk of problems interesting the political scientist concern the actions of such groups. But, from the behavioral perspective, these collectivities exist and behave the way they do only insofar as the people composing them act in certain ways. (1963, pp. 14–15)

This does not mean that in choosing the individual as the unit of analysis, institutions should be of secondary importance for researchers, but that understanding institutions, whether formal – such as governments – or informal – such as economies- or commons-governing institutions (Ostrom, 1998, 2011; Poteete, Janssen, & Ostrom, 2010), should start with the individual. Again, in Eulau's words: "Political institutions are never more or less different from the patterns of behavior of the people who create them or the regularities of their actions. If this be so, institutions can and must be analyzed in terms of the behavior of their molecular units, the individuals whose relations to each other and behavior towards each other are more or less rigidly structured" (1963, p. 15).

Advances in understanding the aggregation of individuals in terms of interpersonal relationships go far in helping us understand the institutions that result. Specifically, Robin Dunbar found a relationship between group size and neo-cortex size in primates and, extrapolating off of this, predicted that humans should live in social groups averaging around 150 (147.8 $\pm$ 45.8) individuals (Barrett et al., 2002). This "Dunbar's number" has been corroborated by findings concerning human social groupings such as Christmas card lists, which average 153.5 per person, and Facebook friend lists (Dunbar, 2012), as well as in the ethnographic literature.

Although Dunbar's number does a good job of describing the cognitive limits on group size, including why groups may start to fission at certain threshold numbers, there still remains the question of the structuring of these groups, both within the group itself as well as the fusion of groups together. Specifically, due to the individual's need for the cognitive capacity to remember the faces and identity of their cohort, but also "integrating and managing information about he constantly changing relationships between individuals within a group" (Zhou, Sornette, Hill, & Dunbar, 2005), evidence that there is a scaling of subgroups of individuals to allow for greater or lesser amounts of information about them to be processed. Zhou and colleagues found evidence for a hierarchically inclusive sequencing of groups at the following numbers of individuals: three to five individuals for a "support clique" in which personal advice and help in times of severe emotional and financial distress might be sought; a "sympathy group" in which an individual has special ties with and has contact with at least once a month; and "bands" in which groups of 30–50 individuals are drawn from the group's average 150 members, albeit in a manner that is unstable in comparison with the previous grouping patterns (Hill & Dunbar, 2003; Zhou et al., 2005). Beyond this, Zhou et al. have identified in the ethnographic literature larger scale groupings of approximately 500 individuals in a "megaband" and 1,000–2,000 members of a tribal unit

(2005, p. 440). However, what is most interesting about these groupings is that there is a preferred scaling ratio of around 3 to their structuring into hierarchies. An example of this grouping can be seen in military organizations, perhaps the earliest in-group/out-group organizational structure developed by humanity, in which the first level, that of squads of 10–15 soldiers, is organized into a platoon comprised of three sections (~35); in turn, companies are made of three to four platoons (~120–150), with battalions made of between three and four companies and support units (~550–800). From here, three regiments or brigades, which are usually made up of three battalions and support units (~2,500+), typically make up a division, with corps, the highest level, made up of two to three divisions (Zhou et al., 2005, p. 442).

The patterns seen here are replicated in a range of organizational structures, suggesting that the absolute size of an organization is not as important as how its social structure is organized to allow for cognitive processing by the human brain. While the rationale for this structural pattern is unknown, whether it is based on cognitive capacity, time budgeting, or other reasons, the pattern holds. Additionally, there is a question as to how these patterns are developed, whether through an expanding network, with each individual interacting with new individuals, or discrete subgroups linked by specific individuals who play a broker role by fusing together populations (Sueur, Jacobs, Amblard, Petit, & King, 2011, p. 709).

In addition to intrinsic limitations to the size and structure of social networks, external factors may also play a role in how many individuals are in a group and how they interact. Networks not only benefit individuals due to the role they play in facilitating cooperation, in which case such activities as communal child care, food sharing, grooming, cooperative hunting, coalition formation, and communal group defense are seen in a range of primate groups (Kasper & Voelkl, 2009), but also benefit individuals within the group through the transmission of information and resources. However, at the same time, social networks can also hasten the spread of pathogens and exacerbate the competition for limited resources (Sueur et al., 2011).

Within the realm of politics, the size and structure of social networks plays an important role in two ways, chiefly through its communicative capacity. The first concerns the transmission of information. Here, information may concern specific cognitive factors or may be the emotional feelings one experiences in relation to a putative leader. Networks play a role in the spread of emotional contagion, as Christakis and Fowler found that happiness spreads dynamically through large social networks, even up to three degrees of separation. In their words, "the social network effect of happiness is multiplicative and asymmetric" (2009, p. 6), with happiness having a stronger effect than unhappiness and with greater numbers of happy "alters" having more of an impact than increased unhappy "alters."

Second, network size and structure influences the ability of parasites and other pathogens to be spread through a population, and hence has an influence on the well-being of a group. This has political implications especially as it appears that the disgust response, which is attenuated in some political groups (namely, the Republican Party in the United States), is adaptive for protecting individuals within a group, and the group itself, by preventing it from coming into contact with outsiders.

Moving beyond "methodological individualism" does not necessarily, as seen above, mean jettisoning the individual as a focal point; on the contrary, it means understanding the individual in social groupings as well as across levels of analysis in manners such as those explored by Peter Corning (2008, 2010, 2011a, 2011b). His "synergism hypothesis" examines the dynamic quality of interrelationships at various levels with allowances for both upward and downward causation – all of which must be understood as particular configurations that give rise to natural selection (1998, p. 150). However, while drawing from multiple disciplines to explore and understand political behavior is to be commended, caution in its use should be taken. In the words of John Crook, regarding the preeminent source of evolutionary insights in his time (1976):

> Much in human ethology and in the inferences to be drawn from animal to human studies remains highly theoretical and, exciting though these ideas may be, caution in their application is needed. The uncritical acceptance of the ideas of the non-scientist Robert Ardrey by many who preferred skilled writing and glib thought to solid academic statement sounds a warning. To know the importance of a neighboring science to one's own requires a reading of the original materials. (1976, p. 274)

Approaches such as this makes temporal assumptions based upon the "here-and-now," whereas politics may also be perceived as a long game that takes place over the course of millennia and is registered in genes. Even the transmission of ideas, that is, memes, through time and spatial dimensions benefits from evolutionary insights (Dawkins, 2006). Advances in measurement, experimentation, and data processing have allowed us greater insight into political behavior overall with the timeframe under consideration ranging from milliseconds to millennia and the units of analysis ranging from the gene to the nation state.

MULTICAUSAL AND MULTIMETHOD APPROACHES TO BIOPOLITICS

The quality of research and the resultant ability to make inferences, draw conclusions, and importantly for political scientists, make policy prescriptions, is based upon the strength of the research designs used to carry out this research. Regardless of its theoretical underpinnings, this research needs to be systematic enough to be replicated. Here, the approach using the UTOS framework (Units, Treatments, Observations, and Settings) from Shadish and colleagues provides a solid starting point (Shadish, Cook, & Campbell, 2002). More specifically, this approach states that "each experiment consists of units that receive the experiences being contrasted, of the treatments themselves, of observations made on the units, and of the settings in which the study is conducted" (p. 19).

The units of analysis, as we have stated previously, should focus on the individual political organism. As stated succinctly by Nico Tinbergen "contempt for simple observation is a lethal trait in any science" (Tinbergen, 1963). However, this still leaves a question of what should be observed and how it should be observed, which themselves are decisions influencing what findings are unearthed. In other words, theory-neutral observation is impossible because all observations contain within them substantive and methodological assumptions, even if driven solely by folk psychology (Shadish et al., 2002). Even with this weakness, which can be seen as addressed by multiple scientific revolutions that change what is observed, and as a result the theories driving research questions (Kuhn, 1996), there is still a reliance on understanding causal relationships in which a cause precedes an effect, the cause is related to the effect, and there are no plausible alternative explanations other than the cause. The difference is that within the life sciences, there is an appreciation of different time frames working at the proximate and ultimate causation levels as well as the understanding of the role that the environment plays interacting with the organism being observed.

As a result of the broad range of evidence and complex theory, researchers need to be at least aware of, if not conversant with, a range of research designs that address the treatments and settings of the UTOS. Specifically, while case studies tend to be used most often in political science due to the nature of the discipline, there is a need for the gold standard of research design – true experiments – to rule out plausible alternative explanations for the cause explaining the effect. Case studies, while informative and providing potential theoretical insights, rely upon correlation and statistical controls for insights. And as has been well established, *correlation does not prove causation* due to causal order often not being characterized in case studies. At the same time, theory building might not take into account potential confounding variables, as well as endogeneity within the model, as there is mutual causation between two variables (Shadish et al., 2002). Within evolutionary theory, this issue becomes attenuated due to the varying time frames evolution operates on the organism as well as the complexity of the interaction between environments and the individual organism.

Thus, while insights may be drawn from case studies, there is always the potential for faulty assumptions to be drawn. This is where true experiments provide their usefulness. Specifically, by randomly assigning individuals to different conditions allows us to test the effect of different treatment manipulations. However, the key term here is "manipulable"; many of the things explored by political scientists with an evolutionary focus are nonmanipulable, such as the personality or attributes of a study participant or environmental surroundings. Therefore, true experiments that consider altered human behavior are rare, such as those carried out by Damasio and colleagues to consider differences between "normal" humans and those with brain damage (Bechara, Damasio, Tranel, & Damasio, 1997; Damasio, 2008) and those considering facial display behavior by comparing blind to sighted judoka (Matsumoto, Keltner, Shiota, O'Sullivan, & Frank, 2010; Matsumoto & Willingham, 2009). As a result, the great majority of experimental studies that attempt to manipulate contextual/environmental information are limited in scope, with concurrent concerns about trade-offs between internal validity, the strength of inferences concerning causal relationships, and external validity, which considers the ability to generalize the findings from a study to multiple different individuals, situations, treatments, and measures.

Ways of addressing this is to consider the use of what Shadish et al. (2002) term analog experiments, which manipulates a cause similar to the cause being studied. For instance, research by James Schubert considering political speechmaking posits that historical events can evoke emotional and cognitive response commensurate with contemporaneous events (Salter, 2002). Natural experiments, in which events such as hurricanes, earthquakes, plaques, drought, etc., that are not necessarily manipulable naturally and randomly occur, also provide a rich territory for research. For instance, the September 11, 2001 terrorist attacks on the World Trade Center and the Pentagon provided a fortuitous opportunity to study the internal correlates of the rally response (Schubert, Stewart, & Curran,

2002). At the very least, quasi-experiments, in which comparison groups are used to allow for inferences to be more accurately drawn, allow for stronger inferences than might not normally be the situation with case studies – even those with statistical controls – by comparing change within and/or between individuals receiving a treatment. All of these different research designs provide the opportunity to engage in research programs that are externally valid, while at the same time increasing internal validity through thoughtful controls and comparisons.

Therefore, researchers testing theories driven by evolutionary insights can be seen as necessarily relying upon a broad range of methods and measures. In turn, these methods and measures, and concurrently, their availability, have an effect on the research design used to collect the data, and the approaches used to analyze the said data. In their attempt to develop a taxonomy of human evolved adaptations, in other words a central clearinghouse for the testing of evolved psychological adaptations in human nature, Balachandran and Glass (2012) suggest using eight lines of evidence to empirically test evolutionary assertions: the theoretical, medical, psychological, genetic, physiological, phylogenetic, hunter-gatherer, and cross-cultural evidence (see also Balachandran, 2011). While the approaches we discuss below are not as inclusive, and are segmented differently, they reflect the state of the art in what is being used by researchers attempting to understand political behavior. Furthermore, as researchers we are well aware of a divide between research that attempts to understand the political processing of individuals that is endogenous to the individual, and thus not easily observed, as well as research that considers indicators that, while more easily observed, are no less rigorous in their demands, as well as insights that may be obtained. These measures may be considered as endogenous, that is, within the person and not easily measured. This includes such approaches as brain scans, hormonal-based studies, research using psychophysiological approaches, and studies that consider genetics. While these studies have great potential to help us in proximate policy and political issues, the theoretical background and insights they provide can be seen as drawing upon and adding to all four of Tinbergen's questions concerning behavior. Therefore, in the interest of respecting the tradition of observing and drawing inferences, then testing those inferences through experimental and quasi-experimental designs, we first consider instances where evolutionary theory has been used from the basis for the observation of political behavior before considering how evolutionary insights into political matters have been used to test determinants of political behavior.

Observable Measures of Political Behavior

One key aspect for having and providing insight concerning political interactions is to understand how information is communicated, both through capacity and intent to act via social signals. An extensive and growing research corpus across a range of academic fields, including social psychology, linguistics, communication, and most pertinently for our purposes, ethology, has explored social signal processing. Ethology, with its evolutionary insights, provides a theoretical and practical basis in understanding how individuals communicate. Specifically, organisms utilize both cues and signals to communicate and apprehend capacity and intent (Maynard-Smith & Harper, 2003; Pantic et al., 2011). Here, cues may be identified as features that can be used to guide future actions, whereas signals are acts or structures that affect the behavior or internal state (which may be seen as behavioral intent) of other organisms. In the first section, we consider height and fluctuating asymmetry in facial characteristics to understand how cues are used to guide social and political interactions. In the second section, we consider how signals can affect the actions of others in political contexts by considering nonverbal behavior, specifically facial display behavior by politicians and vocalic actions in a range of political contexts, including presidential debates, city council meetings, and U.S. Supreme Court oral argument.

Cues

Although there are multiple ways of understanding what "cues" are, Maynard-Smith and Harper in their landmark book "*Animal Signals*" suggest that what sets cues apart from signals is that they are a static "feature of the world, animate or inanimate, that can be used by an animal to guide future action" (2003, p. 3). Here, while cues, like signals, affect the behavior of other animals, cues do not evolve specifically for that purpose; these cues are always "on" and once produced, costs nothing more to express. Furthermore, due to these factors, cues are extremely reliable in providing information about an individual. Specifically, physical attributes can signal quality genetics and behavior, which in turn can influence social and political decisions unbeknownst to the individual making decisions. Understanding the role of these cues is useful when considering political activity, as static phenotypic indicators not only influence social and political outcomes but they likely also reflect societal characteristics. Below, we consider the role of height and fluctuating asymmetry in the facial characteristics of politicians.

Height plays an important role through indicating an individual's stature and size, which in turn provides a short-hand decision cue for opponent strength and power, which in turn influences their attitudes toward the social status quo of political units (Petersen, Sznycer, Sell, Cosmides, & Tooby, 2013). Here, both human and nonhuman social animals (chimps, gorillas, baboons, elephants, reindeer, red deer, birds, fish, ants) show a correlation between physical size and social rank (Boix & Rosenbluth, 2014; Judge & Cable, 2004). In their analysis of sexual dimorphism and its influence on leadership preference, Murray and Murray (2011) note that taller individuals are seen as having higher professional status than those of shorter stature. At the same time, individuals with greater authority are perceived as taller than they truly are with national election winners perceived as taller after the election than before. Although height can provide contemporaneous information about social status within a system, the variation in the height of individuals based upon their status can give us information about the inequality within a system. Specifically, those individuals within a political system that have more status are more likely to be better fed, and as a result be taller and larger (thus perpetuating class differences) while being less prone to health problems such as bone lesions due to anemia, incomplete enamel formation on teeth, loss of bone mass, bone lesions from infections, and stunted growth (Boix & Rosenbluth, 2014).

Functional asymmetry likewise is an important cue used by individuals in the selection of their leaders, whether in the public sphere or in the private sector. Groundbreaking work by James Schubert, who applied insights concerning how the face holds cues to health and parasite loads dealt with during developmental stages, and how it applies to the selection of political leaders was seen in research carried out during the early 2000s, in turn raising a great deal of political ire and controversy when it was covered on the front page of the *Wall Street Journal* (Schubert, Curran, & Strungaru, 2011). Despite this political roadblock, researchers following this agenda have found compelling cross-cultural research about the use of facial cues by followers in the selection of leaders based upon facial characteristics across a range of contexts (Kaplan, Freedman, & Iacoboni, 2007; Livingston & Pearce, 2009; Poutvaara, Jordahl, & Berggren, 2009; Sell et al., 2009; Spisak, Dekker, Krüger, & van Vugt, 2012; Spisak, Homan, Grabo, & Van Vugt, 2012; Todorov, Mandisodza, Goren, & Hall, 2005). Recent work shows that contextual information, specifically situations in which external conflict is being addressed, influences the choice of leaders based upon their facial characteristics (Little, Burriss, Jones, & Roberts, 2007; Spisak, Dekker, et al., 2012; Spisak, Homan, et al., 2012).

Furthermore, evolution-based leadership approaches hypothesize that the masculinity and femininity of a face may matter more than its sex due potentially to properties that enhance signaling of different emotions (Spisak, Nicholson, & van Vugt, 2011). While findings vary depending on the context, the original findings of Schubert have been replicated and built upon by a broad range of researchers.

Signals

Signaling theory arises from both the fields of economics and evolutionary biology. Within the field of economics, the groundbreaking work of Thorstein Veblen at the turn of the twentieth century introduced the concept of "conspicuous consumption" in which individuals signal their wealth to others through the items they buy. In the words of Veblen, this is "specialized consumption of goods as an evidence of pecuniary strength" (2007, p. 43). While this approach to signaling has remained an important part of understanding human and firm behavior for economists and those scholars interested in market interactions (Spence, 2002), it was considered from a predominantly rational actor perspective. Only recently has this approach incorporated a perspective that takes into account evolutionary theory (Saad, 2007).

While economic theory provides a proximate exploration of the influence of cheap talk and costly signals, evolutionary theory, by drawing upon functional premises that work across species, provides theoretical and practical insights that may be applied throughout human interactions. As a result, evolutionary theory's attempt to understand the ultimate causes of behavior is of great value here by providing a unifying framework to work from with the premise that many of the adaptations of species occur not just due to the ability to simply survive but also due to sexual selection, the ability to reproduce at greater rates than competitors. As a result, the complex negotiation of sexual choice and access between males and females is responsible for the development of many personal qualities and their reliable communication through not only cues but also costly signals of physical capacity. As put by Geoffrey Miller "(s)exually selected costly signals typically advertise two classes of traits: good genes or good parenting abilities" (2007, p. 101). Those choosing to mate based upon beneficial (or at least neutral) characteristics can be expected to leave more offspring behind. These offspring will in turn inherit these characteristics. This in turn influences not only our choice of mating partners but may also be a factor in how we choose our leaders, especially as there appears to be a relationship between preferred parenting styles, whether liberal or authoritarian, and political ideology (Hetherington & Weiler, 2009).

Two categories of costly signaling stand out. The first is that of physical fitness in which health, fertility, and youth are signaled through the display of physical attributes as well as indicated through such costly activities such as sports and risk-taking behaviors (Miller, 2000). Jared Diamond argues that costly activities such as smoking, drinking, and recreational drug use provide costly signals of physical quality to others by virtue of their providing a form of self-handicapping. Specifically, profligate consumption of these substances while maintaining control of ones' physical and mental state suggests mastery, in turn leading to the earning of prestige from peers (Diamond, 1992).

The second major category is that of mental fitness in which intelligence and personality are communicated through problem solving and social skills as well as such costly activities as art, music, poetry, and humor (Miller, 2000; Stewart, 2012; Zahavi & Zahavi, 1997). There is obviously an interaction between both physical and mental fitness as physical attributes used in signaling are in and of themselves an indicator of one sort of fitness, while communicating psychological fitness. Specifically, nonverbal displays through the face and through vocalic qualities can not only provide cues of health, youth, and size but they can also encode cognitive messages while communicating behavioral intent.

Nonverbal communication is an incredibly broad category concerning how individuals of many species attempt to influence others through social signals, but can be seen as not inclusive of spoken and written language. Here, facial displays, body language, haptics (touching behavior), proxemics (how close you are to someone), vocalics (what comes out of your mouth that isn't specifically language), and even odors play a role in how individuals respond to others. Those individuals, regardless of their species, that have better production, recognition, and interpretation of these displays can be expected to live longer and potentially thrive, and as a result are more likely to pass along their genes – with their progeny also more likely to be successful. So, in evolutionary terms, capacity in nonverbal communication is beneficial for survival, especially as humans have only recently (in evolutionary terms) developed speech; as a result, nonverbal displays are more automatic and reflect basic needs, values, and appraisals in social situations.

When it comes to humans, we have a preference for and greater accuracy in production and interpretation of emotions based upon the channel it is communicated. For instance, intimate relationship emotions such as love and sympathy are best communicated through touch (haptics), whereas the social status emotions of pride, embarrassment, guilt, and shame are most effectively communicated through the body postural channel. Finally, and most pertinently for a great amount of research concerning political leaders, the survival-based emotions of anger, disgust, fear, happiness, and sadness are best encoded by the face (App, McIntosh, Reed, & Hertenstein, 2011). Here, the "Dartmouth group" model of behavioral leader–follower interactions based on the competitiveness of a behavior and the rank of participants (Masters, Sullivan, Lanzetta, McHugo, & Englis, 1986; Stewart, Salter, & Mehu, 2009) provides an ethologically based approach to understanding facial display behavior. In competitive situations, the dominant individual (i.e., the leader) maintains social order by exhibiting threatening facial behavior such as anger and disgust, whereas subordinates (a leader's followers/subordinates) will exhibit submissive behavior such as fear to signal compliance. In noncompetitive situations, the dominant person will enhance affiliation with the group by reassuring subordinates with facial displays of happiness or neutral affect, while subordinates in noncompetitive situations will express submissiveness through appeasement gestures such as sadness (Salter, 2007). Research observing facial display behavior in the workplace (Salter, 2007), during presidential debates (Stewart, 2012), and television news coverage (Bucy & Grabe, 2007; Grabe & Bucy, 2009) have provided extensive insight into leadership behavior and have further elaborated upon the findings of the Dartmouth Group.

Likewise, verbal and vocalic behavior of leaders have provided insights into local group politics and what may be considered the eminent group political decision body in the United States, the Supreme Court. Schubert, Wiegele, & Hines, (1987), with funding from NIH's National Institute on Aging, undertook a comparative study of the effects of age (an individual property) and age structure (a group-level characteristic) on political decision making in local councils in three states (New York, Illinois, and South Carolina). The study tested multiple hypotheses concerning the degree of influence exerted by older members of local councils and how they exerted this influence. It also sought to determine if older councils (i.e., gerontocratic councils) functioned differently from mixed-age and younger councils. Drawing upon studies by ethologists of dominance structures, they engaged in direct observation of the behavior of councils over a period of one calendar year, observing more than 250 meetings of local councils totaling 500 hours of direct observation time using audio tape recorders, pocket computers, and extensive field notes to collect data. In addition, interviews were conducted with council members (Schubert, 1988; Schubert, Wiegele, & Hines, 1987).

Schubert et al. (1987) summarized their results, stating: "Findings show that old age was associated with greater participation in the group by actors and greater responsiveness by the group toward actors; while with regard to relative age, younger members received higher responsiveness from groups and participated at grand mean levels. Overall, being old was associated with status and involvement in small-group political interactions, but the ascriptive prerogatives of older age did not close out the opportunities for achievements by the relatively young." The researchers also found that older groups proceeded more cautiously and more informally than did younger groups. The more homogeneous the groups (younger or older), the more combative or agonistic the leadership style, but the leadership style was also less oppositional. A particularly important finding that relates individual characteristics and group properties was the following: "We found that chronological age has important, but far from determinative effects on power." Apart from its contribution to role, seniority, and status, elderly age appears to define a special status of social position within groups that might be utilized as a basis for the exercise of influence. While these findings are more than a quarter-century old, they resonate in aging Western societies where older populations control economic and sociopolitical resources and as a result have considerable sway over policy outcomes.

Research carried out by James Schubert and colleagues attempted to infer why justices act in the manner they do through analysis of nonverbal cues that communicate emotional state (Schubert, Peterson, Schubert, & Wasby, 1992; Wasby, Peterson, Schubert, & Schubert, 1992). Their research used a range of linguistic and paralinguistic indicators to examine oral argument in their extensive and time-intensive inquiry. One of their indicators, interruptions, provided a highly salient, face valid, and recognizable indicator of attempts to gain dominance, and at the same time reduce the status and power of a target, and has been used in recent studies building upon these findings to understand the current Supreme Court and its decision-making process (Johnson, Black, Goldman, & Treul, 2009; Johnson, Black, & Wedeking, 2009; Kimmel, Stewart, & Schreckhise, 2012).

Measures of Political Behavior That Are Not Easily Observed

Although observational studies of political behavior can be tedious, they provide insights that advance our understanding of social and political behavior. However, testing these insights remains an important next step to progressive scientific understanding, which in turn is premised upon reliable and valid measures. In this section, we introduce and discuss different measures currently being used to understand the influence of biology on political behavior, starting with a psychophysiological approach.

Measuring response to political stimuli through what may be termed psychophysiological means provides two beneficial services. The first is that it provides insight into the dispositions and orientations held by individuals that might be otherwise difficult to assess, especially when these responses might not be seen as socially acceptable. In the words of Kevin Smith and John Hibbing "(T)he primary contribution of a psychophysiological approach, though, is not simply as another measurement technique but as a means to more directly and intuitively connect biology and political attitudes and behaviors" (2011, p. 229). The second service this approach provides is to understand connections between response to proximate stimuli and more ultimate explanations, especially as the neural systems involved in these physiological responses are evolutionarily old and shared with mammalian species (Gray, 1987; Panksepp, 1998).

The measurement of human physiology and linking it to psychological processes can be seen as having two major divisions. The first is the central nervous system (CNS), which consists of the brain and the spinal cord. The second is the peripheral nervous system, which connects the CNS to the organs and the other parts of the body. In turn, the peripheral nervous system can be seen as having two subsystems, the somatic nervous system in which there is a level of conscious control and the autonomic nervous system (ANS) which operates largely outside conscious awareness. Below, we discuss how the CNS and the peripheral nervous system are being considered in order to understand political behavior by first analyzing how different technologies are used in brain scans, and second by looking at political insights provided by the ANS by various different measurement tools.

The Central Nervous System and Brain Scans

Analysis of the CNS has traditionally been carried out through extrapolation from animal studies (Gray, 1987; Panksepp, 1998), insights from autopsy, and inferences from trauma and disease (Damasio, 2008). While this structural approach has been of great use, its static nature limits insights. Newer approaches that couple increasingly sophisticated sensing devices with powerful computing methodologies has led to techniques that can be noninvasive and safe while at the same time provide groundbreaking insights into the brain and individual behavior. Here, brain activity is measured through two different means, electrical activity and blood flow. The former is based upon electrophysiological principles that consider changes in baseline activity in specific regions of the brain to understand cognition, emotion, and behavior. The tools used here include electroencephalography (EEG), magnetoencephalography (MEG), and transcranial magnetic stimulation (TMS).

The second means by which brain activity is measured, and inferences about psychological processes are made, is through movement of substances in the blood that nourish and replenish active areas. For instance, positron emission tomography (PET) scans create images mapping how radioactive substances, ingested by the participant, are distributed in the brain to measure such things as blood flow, oxygen consumption, glucose utilization, and neurochemical activity. The more readily available, and arguably more powerful, functional magnetic resonance imaging (fMRI) technology – due to its greater resolution and detail – relies on blood-oxygenation-level-dependent (BOLD) contrasts that consider blood flow to a brain region. This in turn is premised upon measuring the magnetic qualities of iron-rich hemoglobin that floods into the metabolically depleted area of the brain under consideration (Blank, 2013).

With continuing advances in brain imaging techniques, there has been increasing insights into the functional anatomy of the brain and how different parts of the brain contribute to different types of thinking and reasoning. Perhaps, most tellingly, these insights about both structural and functional neuroscience have in turn influenced the way that we think about thinking. Whereas previously rational thought was at the forefront of understanding human decision making, now with insights from brain imaging techniques applied to social neuroscience, we have a greater appreciation for the interplay of preconscious processing with the construction of social and physical reality in the consciousness. In other words, researching the brain helps us understand the mind. Specifically, we have a greater appreciation for the fact that we are continually processing information from the sensory apparatus of the eyes, tongue, ears, olfactory bulb, and skin, while monitoring somatosensory information from the muscles, skin, joints, and other various nerve receptors that tells us information about aspects of the body, including temperature, position, movement, and difficulty (Marcus, 2013). The sheer volume of information processed on an ongoing basis suggests that consciousness is only a small sliver of brain processing.

At the same time, the insights from the brain sciences suggest that preconscious processing is much faster than previously thought. Whereas we are capable of sustaining the illusion of movement when watching movies and televised images presented at 24 frames per second, with images presented at 1/33rd of a second being seen as "subliminal" (Stewart, 2008; Weinberger & Westen, 2008), processing of information occurs well before then. Specifically, while conscious processing is seen as occurring at 500 milliseconds and preconscious processing occurs at between 50 and 100 milliseconds (Marcus, 2013), recent research suggests that awareness for some individuals can occur at 13 milliseconds.

In addition to insights into the mass and speed of processing, brain imaging techniques has provided us insights into specific responses to stimuli with political implications (Schreiber, 2011). However, care must be taken with the application of these insights to social, and especially, political issues. Martha Farah, in her review and response to critiques of functional neuroimaging, finds kernels of truth in four major areas of criticism, which should be considered and applied when reading and carrying out this type of research (2014). The first criticism, which points out that fMRI does not consider actual neural activity but an indirect indicator (BOLD) and also relies upon methodological assumptions concerning measurement and data representation, can be applied to most scientific endeavors, including biopolitical research attempting to understand political behavior. The second critique concerning metatheoretical assumptions and goals, however, is of particular interest to those hoping to explain human behavior. Specifically, the focus on localizing specific brain activity and tying it to psychological theory, which in turn often treats the brain as modular, might misdirect researcher focus. And perhaps more pertinently, this approach may lead to *reverse inference* in which this technology may be "used wantonly, interpreting a pattern of activation without knowing, or acknowledging, the variety of psychological processes that could produce that pattern." (S24). The third critique is one that may be applied to all "big data" in which "p-hacking" using large numbers of independent statistical tests leads to significant results due to chance. Additionally, the selection of subsets of the population to focus on may lead to corroboration of hypotheses more so than correctly and fairly testing the null hypothesis that there is no effect. Finally, and perhaps most pertinent for research that has direct implications for social and political policy, the images of brain action that are produced are visual fictions that may be overly compelling and convincing, even when the science itself may not be so persuasive (Farah, 2014).

Despite these cautions, which may be seen as elaborations on how science should be undertaken – with the full awareness that there are no privileged data sources, status, beliefs, or findings, and there is no closure (Marcus, 2013) – findings in the brain sciences concerning neural activity suggest different functions played by different parts of the brain. For instance, the fusiform face area (FFA) of the fusiform gyrus is responsible for processing basic information about faces and potentially information about gender characteristics, as well as ethnicity, and may be seen in how stereotypes are processed, especially in politics where facial characteristics are important for the selection of politicians as representatives (Todorov et al., 2005).

As will be discussed in more detail in Chapter 5, the amygdala is involved with processing fear as well as other negative emotions such as disgust, sadness, and anger and is also involved with positive emotions (including those dealing with humor and sexuality) and implicit racial attitudes. This is highly important when responding to potential threats to well-being and ascertaining strategies to deal with them, whether they come from external foes, internal threats, or natural disasters. Research suggests differences in the activation of this portion of the brain in Democrats and Republicans, with the latter showing greater activity in this region when engaged in risky decision making (Schreiber et al., 2013). Interestingly, findings suggest that explicit attention is less likely to activate the amygdala (Schreiber, 2011).

Awareness concerning how other parts of the brain are implicated in social and political activity is likewise being drawn. For instance, the anterior insula is involved with perception of internal physiological states, as well as with empathy (not just pain, but also with social exclusion), and is associated with higher levels of activation in Democrats in risky decision making (Schreiber et al., 2013). This region and the emotional processing that occur (Petersen, 2010; Petersen, Sznycer, Cosmides, & Tooby, 2012; Petersen et al., 2013) may, thus, influence the types of social welfare policies engaged in, as well as who is involved and how they are involved. The basal ganglia (including ventral striatum, globus pallidus, and substantia nigra) is associated with valuation, reward, and decision making, and thus could influence how we understand negotiation. Likewise, the anterior cingulate cortex (ACC) is involved with processing decisions and potential conflicts, especially when detecting other people's errors. Finally, the medial prefrontal complex (mPFC) is understood to be important for social cognition – as seen in the case of Phineas Gage – as well as a range of moral and ethical judgments, and has provided an important entry point for understanding how the brain influences behavior (Damasio, 2008; Ekman & Davidson, 1994).

Measuring the Autonomic Nervous System through Psychophysiology

The ANS, which operates mainly outside of conscious awareness, can be divided into two subsystems which help us understand human, and hence political, behavior. The first is the sympathetic system which is mobilized in response to threats and stressors as well as arousal more generally. This subsystem is measured through increased heart rate and blood pressure as well as blood being moved to the muscles under voluntary control. In contrast, the parasympathetic system is involved with energy conservation and resource restoration by such actions as decreasing heart rate and increasing digestion (Smith & Hibbing, 2011). What makes ANS measures particularly interesting is their ability to measure physiological activities well before cognition occurs (Bucy & Bradley, 2010; Marcus, 2013). We discuss a few of the most often used psychophysiological measures below.

Electrodermal activity (EDA) is measured primarily when considering variation in skin conductivity, mainly in the hands. Here, eccrine sweat glands are innervated in preparation for the body to take action, with perspiration approaching the surface of the skin and measured through electrical currents considering skin conductance levels (SCLs) (Bucy & Bradley, 2010; Smith & Hibbing, 2011). Cardiac measures are used to consider ANS activity through measuring the heart rate, blood pressure, and respiration of an individual. Here, the ANS regulates through both sympathetic and parasympathetic measures through the contraction and relaxation of the heart's muscle fibers in which the timing and variability of the heart pump (chronotropy – measured through the electrocardiogram [ECG]) and how hard and how much blood the heart is pumping (ionotropy). Respiration is considered through the breathing rate and amplitude and can be measured through relatively cheap approaches such as bellows (Bucy & Bradley, 2010; Smith & Hibbing, 2011). Finally, electromyography (EMG) is used to measure electrical activity over muscle groups and can be seen in a range of different sites. While EMG has been used to measure "gut activity" with electrode sensors placed on the stomach to consider disgust reactions, most placements occur on the face where communicative signaling occurs most often. The first major site is the orbicularis oculi, the muscles surrounding the eye, to consider eyeblink startle reflex, which occurs in response to aversive stimuli (Smith & Hibbing, 2011). Other popular sites to measure response is the zygomaticus major, which is seen as measuring positive emotional state due to its involvement with the lip corner pull involved with smiles; the corrugator supercilii, which is the muscle in the middle of the forehead that is contracted during frowns and other negative facial displays; as well as other varied sites such as the orbicularis oris which surrounds the mouth and is used to control facial mouth movements; the levator labii, which pulls the upper lip upward as is associated with disgust; as well as a host of other potential sites on the body.

While these psychophysiological measures of the ANS are relatively rarely used by political scientists, they have been used by enterprising early biobehavioral scholars valuing their ability to measure response to political stimuli generally (Lodge, Cross, Tursky, & Tanenhaus, 1975; Wahlke & Lodge, 1972) and, more specifically, in response to facial displays by political leaders (McHugo, Lanzetta, & Bush, 1991; McHugo, Lanzetta, Sullivan, Masters, & Englis, 1985). More recent psychophysiological studies have been carried out by political communication scholars to better understand response to display behavior inappropriate for accompanying news stories (Bucy & Bradley, 2004) and incivility during televised debates (Mutz, 2007; Mutz & Reeves, 2005). Additionally, research using these measures suggests that there are significant differences in response to stimuli based upon political attitudes (Hibbing et al., 2013; Oxley et al., 2008; Smith, Oxley, Hibbing, Alford, & Hibbing, 2011a, 2011b).

Hormones in the Expression and Development of Sociopolitical Response
Although not necessarily as immediately responsive to external political stimuli, a range of hormones have been shown to play an integral role in reflecting and structuring behavior with power and political implications, as well as affecting developmental processes that impact how individuals vary in their response to stimuli (McDermott, 2011a, 2011b). In terms of analyzing social behaviors and the influence of context, a range of hormones have provided insight into behavior related to politics including testosterone, cortisol, oxytocin, vasopressin, and monoamine oxidase A (MAOA). Testosterone is the most prevalent of five androgens produced in the male testes, female ovaries, and adrenal cortex, and is not only involved with regulation of sexual desire but also prototypically male behaviors (McDermott, 2011a, 2011b). Likely as a result, it is also highly studied with findings suggesting changes during a person's life cycle and in response to specific events such as sporting and chess competitions, with testosterone higher postevent for winners than for losers (Madsen, 1986; Mazur, 2005). Likewise, strong in-group identifiers with sports teams (Mazur, 2005) and presidential candidates (Apicella & Cesarini, 2011) experience increased or decreased testosterone levels in response to winning or losing, respectively.

Cortisol is involved with response to internal and external stressors, elevating in response to these stressors, and as such plays a highly important role in regulating social dynamics. Although popular accounts suggest that leadership is stressful, research considering cortisol levels suggest those in these positions have lower levels than comparable nonleaders (Sherman et al., 2012), which likely leads to better health outcomes (Marmot, 2004). The neuropeptides vasopressin and oxytocin have been shown as playing a role in prosocial affiliation and social recognition and attachment, with the latter affecting social interaction and sensitivity to rewards through the reduction of anxiety and stress (Bethlehem, Baron-Cohen, van Honk, Auyeung, & Bos, 2014). Finally, MAOA is shown to be connected with impulsivity and propensity for violence (McDermott, 2011a, 2011b). Although all these hormones are expressed in response to internal (endogenous) and external (exogenous) factors, the likelihood and levels expression, and concomitant behavior, are influenced by an individual's inherited genetics, which we explore next.

Genetics

A major goal in biopolitical research is to understand the role of inherited genetics on political behavior. While there is a great deal of concern and criticism about the equation of genes with an individual's destiny, which may be seen as warranted given the historical misinterpretation and abuse of scientific knowledge in the name of Social Darwinism in the service of the eugenics movement and ethnic cleansing, understanding the role of genetic variation on human behavior has implications for the proper modeling of political activities and how to best construct public policies. Research in the area consists of two distinct, but overlapping, behavioral genetic techniques: twin and kinship models and molecular genetic analyses. Twin and kinship designs allow for population estimates of the amount of variation due to genetic and environmental factors and molecular genetic approaches identify specific genetic variants related to the trait of interest. Here, identical (monozygotic) twins are the gold standard in natural experiments allowing for disentangling the effect of biological and environmental influences on variation in humans, although studies of fraternal (dizygotic) twins and nuclear families have been useful in teasing out correlates of political behavior (Eaves et al., 2011; Hatemi, Medland, Morley, Heath, & Martin, 2007).

Not surprisingly, twin studies have drawn sharp criticism. For instance, Shultziner (2013) argues that they do not test the hypothesis that they set to prove, namely, that there is a genetic effect on political attitudes, but instead merely show that trait similarity is accounted for because identical genomes tend to react similarly to a similar environment. Moreover, the measurements in twin studies exaggerate the genetic effect and confound the genetic and environmental effects. The combination of the two problems yields inflated heritability estimates and false-positive results. In twin studies, the heritability measurements for political preferences (many over 0.4) are often more than twice those of many medical disorders and other physiological traits (many lower than 0.2). There is a wide consensus, even among scholars working to integrate biology and political science, that political traits are clearly affected by multiple genes (polygenic traits). Thus, for Shultziner, there is clearly something suspicious about the fact that heritability measures for political traits are greater than the expression of diseases.

The second technique is the use of molecular DNA to identify the specific genetic markers that underlie the genetic variance from heritability estimates; in other words, candidate-gene-association (CGA) studies. Although to date, few genetic markers have been closely associated with behavioral outcomes and specific traits in the predominantly "candidate-gene" studies that rely on identifying and testing for specific genes. This is likely due to the complexity of gene–environment interactions, as well as interactions between the genes (epistasis) themselves. Despite this, advances in technology, methodology, and data availability, including genome-wide association studies, hold promise that pathways from genes to complex political behaviors and preferences will be identified.

The use of CGA studies in the social sciences has been facilitated both by the increasing availability of large data sets that contain both largely self-reported behavioral data and limited genetic data. In most CGA studies carried out in political science, a researcher proposes that a gene variant will predict a self-reported behavior. It is a foundational assumption of CGA studies in the social and behavioral sciences that one or two common polymorphisms can so greatly increase the likelihood of a complex behavior as to predict that behavior for a defined population, if not for an individual. The use of CGA studies to study political behavior and attitudes has yielded enticing, yet highly controversial results. A particular gene (or gene variant) has been claimed to predict salient social activity such as punishing behavior in an experimental setting (McDermott, Tingley, Cowden, Frazzetto, & Johnson, 2009), cooperative behavior in a trust game (Cesarini et al., 2009), and preferences for the voluntary provision of social goods (Mertins, Schote, Hoffeld, Griessmair, & Meyer, 2011). In the arena of political activity, research by Alford, Funk, and Hibbing suggests that when it comes to political ideology (specifically, whether one is a liberal or a conservative), genes count for more than environment (2005).

In response to a 2008 *APSR* article by Fowler, Baker, and Dawes on genes and voting, Charney and English published an article highly dismissive of all CGA studies. They argued that the number of recent claims by social scientists to have discovered a statistically significant association between a particular common gene variant, either on its own or in conjunction with a particular environment, and politically relevant behaviors warrants careful evaluation. They concluded: "we have the strongest reasons to doubt that a handful of candidate genes will provide a meaningful key to understanding differences in voting behavior, political ideology, or attitudes toward abortion. Advances in genomic science suggest that this reductionist approach is ill conceived, and proponents of genopolitics have yet to present compelling evidence to the contrary" (2012, p. 30).

In 2013, Fowler and Dawes responded by replicating part of their earlier study and arguing that many of the Charney and English characterizations of the literature in genetics and political science were misleading or incorrect. As noted in Chapter 2, in a separate *APSR* article, Deppe, Stoltenberg, Smith, and Hibbing (2013) confirmed the original conclusions of Fowler and Dawes, but also supported several of the critiques suggested by Charney and English. In the latest installment, Charney and English fired back:

> … from the standpoints of genetics, neuroscience, and evolutionary biology, genopolitics is a fundamentally misguided undertaking; we also respond to FD's charge that some of our previous statements concerning genetics are "highly misleading," "extremely disingenuous" and "even incorrect." We show that their criticisms demonstrate a lack of awareness of some basic principles in genetics and of discoveries in molecular genetics over the past 50 years. (2013, p. 382)

Although dealing with issues that may be seen as distinctly genetic, such as epigenetics where the expression of genes is influenced by the interaction of environmental with genetic factors, much of the critiques reflect those in the brain sciences. Specifically, questions arise concerning not only measurement and statistical testing of phenomena but also the appropriateness of using such tools to understand political behavior, indeed any social behavior, due to there being levels of analysis problems (Charney & English, 2013).

PERSONALITY, POLITICS, AND CHANGING INTERPRETATIONS AND INDICATORS

Of long-standing interest to those wishing to understand human political behavior is the belief in a predictable human nature, in other words a stable personality trait. Personality traits can in turn be seen as setting relatively stable parameters around behavior, and is the result of complex interactions between genetics and environmental factors. These personality traits allow for prediction concerning how individuals would act across different contexts, allowing for long-term social connections to be made.

These personality traits in turn influence what is seen as the "best" type of government to help structure social interactions. While the influence of philosophical conceptualizations of humanity and their personality traits on governance can be traced back to Plato's Republic in ancient Greece, with his analogy between elements of the soul and that of the city being represented by the appetitive producers, spirited soldiers, and rational guardians (Neblo, 2007), more recent conceptualizations of the fundamental tenets of man, as existing prior to society in the "state of nature," by Thomas Hobbes, John Locke, and Jean Jacques Rousseau have shaped and were shaped by changing society during the enlightenment (Masters, 1989). Indeed, the American form of government, while drawing upon these philosophers, established the separation of powers and federalist structure based upon a realistic and sophisticated understanding of human nature by the authors of the Federalist Papers who took into account the tendency for political actors to form factions (Peterson & Somit, 1989).

Although there is indeed a political "human nature," questions remain as to whether and what are the types and/or dimensions of individual differences in this human nature, how stable they are, and how this influences political behavior. Concurrently, there are questions of measurement which influence all the foregoing. Historically, and likely due to ease of application, approaches to individual differences in personality and temperament have focused on typologies to define individuals. The ancient tradition of Egypt, Greece, and Rome was the four-fold typology of the humors that focused on the animating fluids of black and yellow bile, phlegm, and blood with the "dominant fluid" leading to the specific personality types of melancholic (sad and depressed), choleric (angry, aggressive, and ill-tempered), phlegmatic (calm, steady, and imperturbable), and sanguine (courageous and desirous), respectively (Marcus, 2013). More recently, the Myers–Briggs test, with 16 different personality types, and the right-brained/left-brained typologies have dominated popular discourse, albeit without extensive research support for assertions made by advocates.

Personality has been shown to have substantial contributions both from genetics and from environmental adaptations. The predominant means by which personality is currently defined and approached by researchers is through a broad set of traits known as the Big Five (Costa & McCrae,

2008; Goldberg, 1990), which in turn built off of Hans Eysenck's two traits of extraversion and neuroticism (Marcus, 2013). The traits of Big Five are known by the acronym "OCEAN": openness to experience, conscientiousness, extraversion, agreeableness, and neuroticism (also known as emotional stability). Openness to experience implies the traits of imaginative insightfulness, having a wide range of interests, complexity, and openness to life. Conscientiousness is composed of the traits being well organized, planning, thorough, and careful. Extraversion includes the behavioral traits of being talkative, assertive, energetic, and expressive. Agreeableness is comprised of the traits of kindness, affection, caring, and sympathy. Neuroticism, or its inverse – emotional stability – includes the traits of moody, tense, anxious, emotionally unstable, irritable, and filled with sadness. While these traits may change over the course of an individual's lifespan, they are relatively stable and provide insight into a range of behavioral strategies, an insight increasingly applied by political scientists (Gerber, Huber, Doherty, Dowling, & Ha, 2010; Mondak, Hibbing, Canache, Seligson, & Anderson, 2010; Weinschenk, 2014). Further, while most researchers measure the Big Five through traditional self-report formats, an alternative approach by Markus Koppensteiner has considered measurement by observation of body movements during speechmaking activities with some success (Koppensteiner, 2011, 2013; Koppensteiner & Grammer, 2010, 2011). Indeed, thin slice judgments of personality, as measured by the Big Five, may be made based upon short video clips (Borkenau, Mauer, Riemann, Spinath, & Angleitner, 2004) and even the offices and bedrooms an individual occupies (Gosling, Ko, Mannarelli, & Morris, 2002).

Research has also considered the relationship between the Big Five and environmental and genetic influences (Eaves et al., 2011), as well as how they influence consumption and other decisions and actions (Miller, 2009; Stewart, 2012). In sum, this measure of personality traits has seen an increasing relevance in explaining political behavior, including voting (Hibbing et al., 2013; Tuschman, 2013), although care must be taken in interpretation of the results. For instance, while Wolak and Marcus found that the Big Five personality traits of openness to experience, extroversion, and neuroticism play a role in emotional response to political stimuli, their effect on strategic and tactical political response is mixed (2007).

However, personality is not necessarily limited to the Big Five; other approaches with use response tendencies toward using specific emotions provide insights into individual response, as well as environmental influences on these individuals. The sensation seeking scale, which measures willingness to take risks, has been useful for understanding certain types of liberalism (Jost, Glaser, Kruglanski, & Sulloway, 2003; Zuckerman, 1994), whereas the positive affect negative affect scale (PANAS), while a broad measure, has been quite useful in measuring temperament (Watson, Clark, & Tellegen, 1988).

More finely grained emotional response tendencies, such as anxiety, anger, and disgust, have been tied with political attitudes and behavior. For instance, both trait fear and trait anger can be seen as important for understanding in-group/out-group divisions through aggressive confrontation (Hibbing et al., 2013; Skitka, Bauman, & Mullen, 2004). Until recently, one relatively unexplored emotion that is proving quite fruitful in understanding individual political decisions and behaviors is that of disgust. Recent theoretical and psychometric analyses have suggested that state and trait disgust is more than a singular defense system for protecting the body from oral incorporation of pathogens (Haidt, 2012). Cultural psychologists posit that a major reason for cultural adaptations is to respond to environmental threats. One such threat, which will be the focus of the present manuscript, is infectious disease. The resultant behavioral immune system (BIS) leads to greater collectivism with sharper psychological boundaries between in-groups and out-groups (Schaller & Park, 2011). Therefore, at the same time there is stability within organizations; boundaries are set to keep outsiders at distance through obedience to socially conservative value systems that focus on "adherence to social norms, in-group cohesion and out-group avoidance" (Terrizzi, Shook, & McDaniel, 2012). Specifically, the BIS may be seen as operating on individuals through belief and behavior promulgating social organizations.

Tybur and colleagues have developed what they refer to as the "three domains of disgust," which are three clusters of situations or contexts in which the basic emotion of disgust operates to facilitate cultural and individual adaptation and ultimately promote survival and reproductive success.

Theoretical and psychometric analyses support three domains of disgust proposed by Tybur and colleagues, including pathogen (aversions toward pathogen-carrying stimuli or contexts), sexual (aversions toward sexual practices that are less safe or confer less reproductive success), and moral (aversions toward social transgressions) domains of disgust (Olatunji et al., 2012; Tybur, Lieberman, & Griskevicius, 2009; Tybur, Lieberman, Kurzban, & DeScioli, 2012). Moreover, these three systems of disgust have been demonstrated by fMRI studies to be separate, albeit overlapping, domains (Schaich Borg, Lieberman, & Kiehl, 2008), which in turn regulate decisions through information processing mechanisms appropriate for weighting trade-offs in each of these domains (Tybur et al., 2012).

Pathogen disgust likely mediates the rejection of out-groups that potentially pose the threat of contamination (Schaller & Murray, 2011). For instance, beliefs concerning purity may be related to negative attitudes toward out-groups, including homosexuals (Inbar, Pizarro, Iyer, & Haidt, 2012; Olatunji, 2008; Terrizzi, Shook, & Ventis, 2010), foreigners, and immigrants (Hodson & Costello, 2007; Navarrete & Fessler, 2006), and may stem from concern that these groups may potentially transmit pathogens and disease (Tybur, Merriman, Caldwell Hooper, McDonald, & Navarrete, 2010). This in turn limits the potential for infection by pathogens coming from outsiders (Fincher & Thornhill, 2008).

Within social organizations, such as religions, sexual disgust may be emphasized to avoid mating with individuals that may jeopardize reproductive success (Tybur et al., 2012). This reproductive approach tends to champion high fertility, low promiscuity strategies (Li, Cohen, Weeden, & Kenrick, 2010; Weeden, Cohen, & Kenrick, 2008) by setting, monitoring, and enforcing social group norms and values by "avoiding reproductively costly sexual behaviors, narrowing the pool of sexual behaviors and partners to those likely to contribute to the production of healthy viable offspring" (Tybur et al., 2009, p. 106). At the same time, sexuality-specific avoidance due to disgust allows for a range of beneficial social interactions that might have been precluded by pathogen-based disgust (Borg & de Jong, 2012) while also limiting potentially reproductively costly within-group behavior. Moral disgust likewise plays an important social role by limiting potentially maladaptive behaviors that disrupt relationships and group cohesion (Chapman & Anderson, 2013; Russell & Giner-Sorolla, 2013; Tybur et al., 2012). Moral disgust may be seen as most divorced from pathogen avoidance and response, yet utilizes many of the same physiological, psychological, and behavioral responses as pathogen and sexual disgust.

In conclusion, personality traits serve many different purposes in social and natural environments and may be measured and conceptualized in many different ways. Whether measured through cues and signals or other not easily observed measures such as brain scans and even genetic markers (Eaves et al., 2011; Wolak & Marcus, 2007), or even traditional self-report approaches, understanding individual traits and their presence and variation in different environments can provide useful insights into political behavior. Specifically, it might help scholars predict and respond to leaders of various groups as well as understand the partisan identity, policy preferences, and potential for political participation within and between groups of individuals.

CONCLUSIONS

The broad range of data collection methods and variety of research designs that may be used in the implementation of inquiries provides not only the potential for greater precision in inferences drawn, but also the possibility for enhanced theoretical advances. Although the diversity of approaches explored in this chapter can be seen as potentially overwhelming for a single researcher hoping to master a single or even a subset of method(s), being aware of the broad range of options and concomitantly opening cooperative connections across disciplinary boundaries can only serve to benefit social and political inquiry. Specifically, the ability for researchers to triangulate their findings through different techniques and designs allows for more robust conclusions to be drawn as to the nature of human behavior. It is to these findings that we focus our attention in the next chapter on comparative politics.

CHAPTER 4

COMPARATIVE POLITICS, WORLD POLITICS, AND INTERNATIONAL RELATIONS IN BIOPOLITICAL PERSPECTIVE

Samuel M. Hines, Jr.

Almost unremitting dramatic events in a highly volatile world politics raise many critical issues for political and social scientists. These events and the interpretative and explanatory challenges they present confirm the continued importance, indeed imperative, of creating a discipline of political science that takes the life sciences seriously, particularly evolutionary science, genetics, cognitive neuroscience, environmental science, and ecology as well as the study of primate and human ethology. Fortunately, over the last decade, significant and transforming scholarly work has emerged that can now begin to better inform the study of comparative and world politics. This body of work includes, but is not limited to:

- major new synthesizing interdisciplinary works by Corning (2003, 2005), Diamond (1997, 2005), Fukuyama (1999, 2004, 2011), and Sanderson (2001, 2008, 2014) that challenge the standard social science model (SSSM) and advance biosocial science based on an evolutionary model (EM);

Politics and the Life Sciences: The State of the Discipline
Research in Biopolitics, Volume 12, 101–124

ISSN: 2042-9940/doi:10.1108/S2042-994020140000012003

- advances in the application of evolutionary theory to the study of international relations, especially work in evolutionary world politics by Modelski (2001), Thompson (2001), Thayer (2004), and others (see http://faculty.washington.edu/modelski/);
- explicitly biopolitical research programs of Falger (1991, 1994, 1997, 2001a, 2001b), Thayer (2004), Van der Dennen (1995), and Vanhanen (1992) that investigate the relationships among ethnicity, ethnocentrism, group conflict, and forms of warfare and the correlates of and prospects for sustainable democratic political systems;
- sustained efforts by Peterson and Somit through their *Research in Biopolitics* book series to promote the study of politics informed by the life sciences and by current research in biopolitics;
- public policy research that incorporates findings from the life sciences, especially genetics and neuroscience, and their policy implications (for examples, see Blank, 1981, 1990, 1999, and Funke in this volume), as well as research on environmental politics and policy that informs international relations research which is discussed in Chapter 7 of this volume;
- significant studies on the evolutionary/biological bases of moral and ethical behavior that point to the potential for human cooperation by Gazzaniga (2005), Wright (1994), and de Waal (2006);
- a multitude of relevant articles and reviews in the Association for Politics and the Life Sciences journal, *Politics and the Life Sciences*, since 1982;
- Conferences and publications resulting from the work of Research Committee #12, Biology and Politics, of the International Political Science Association (see Somit & Peterson, 2013b).

Collectively, these developments represent significant advances in our understanding of the importance of modern evolutionary theory and processes, for example, the evolutionary model, human biogrammar (see Box 1), and evolutionary adaptationist strategies, all of which ought to be informative to the study of comparative/world politics and international relations. The careful and persuasive incorporation of life sciences theories, knowledge, and methods into the theoretical formulations of biopolitics by Corning, Falger, Masters, Peterson, Sanderson, Somit, Thayer, and others demonstrates the appropriate use of theories, concepts, methods, and data drawn from the life sciences to achieve an interactive model that properly relates nature and nurture. As discussed in Chapters 1 and 2, this model must be founded on a comprehensive body of knowledge that reveals the power of analytical, holistic, and historical/narrative reasoning to

Box 1. Biogrammars.

Tiger and Fox (1971) argued that humans come equipped with a biogrammar, a basic set of genetically based programs or biological templates that predispose human behavior along certain lines. The predispositions suggested by these authors and others include aggression, hierarchy, male dominance, mother–infant bonding, territoriality, and incest avoidance. All have been controversial. For instance, Tiger and Fox contend that men and women have different biogrammars that dispose them to behave in distinctive ways. The male biogrammar predisposes men to be aggressive and dominant, while the female biogrammar predisposes women to have children and to care for them. Male and female biogrammars are not absolutely fixed as they can change very slowly.

demonstrate how biological and sociocultural evolution can be combined to reflect how both biological/chemical and environmentally shaped factors influence human political and sociocultural behavior resulting in an understanding of humans as both *of* and *in* nature. The enduring features of politics and political behavior, anticipated as early as in Aristotle's characterization of man as a social–political animal, include those features of human life embedded in our human biogrammar (dominance hierarchies, kinship, reproductive strategies, the capacity for selfish and altruistic behavior to include the potential for both conflict and cooperation, the making and use of tools and symbols, and others) continue to shape our political responses both individually and collectively. Versions of the life-science-based interactive paradigm that moves beyond the SSSM (see Fig. 1 in Chapter 1) that were provided by Masters and Schubert (see Fig. 2 in Chapter 1 and Fig. 1 in Chapter 2) some time ago clearly indicate the robustness of this approach.

Earlier analyses of these major fields (comparative and international politics) in political science largely has stood the test of time over a decade (Blank & Hines, 2004; Somit, Seo, & Peterson, 1994), but the discipline of political science, except by some scholars in the area of world politics/international relations, has still not responded to the call for a discipline that takes the life sciences seriously. The frustrations experienced by Somit and Peterson (2005) in finding an audience for their work documenting and explaining the failure of democracy in recent times remains for those endeavoring to promote a life-science-informed political science. As described in Chapter 2, elsewhere in the social sciences, there has been greater impact from evolutionary social science research (Niedenzu, Meleghy, & Meyer, 2008; Sanderson, 2001, 2008, 2014; Turner, 1998; Walsh, 1995).

For instance, the discipline of sociology has seen a number of advances, particularly through the work of Sanderson (2008) and, earlier, Pierre Van den Berghe (1981), resulting from the incorporation of a life sciences perspective. Anthropology, of course, has long been sensitive to, although its practitioners are not always in agreement with the impact of evolution and the theoretical framework that has developed around the Darwinian Conflict Theory (DCT) (see Thayer, 2004). The field of evolutionary psychology has perhaps had the most influence among the social and behavioral sciences (Barkow, Cosmides, & Tooby, 1992; Tooby & Cosmides, 1988; Wright, 1994). As noted earlier, as these life-science-based approaches proliferate and their research findings become more compelling, the SSSM is increasingly called into question. One of the most succinct summaries of the shortcomings of the SSSM and the case for an evolutionary model can be found in Somit and Peterson (2003b). Similar comparisons can be found in most of the major arguments on behalf of the EM that are included in our extensive bibliography.

COALESCENCE OF FIELDS AND SUBDISCIPLINES

The fields of comparative politics and international politics afford a number of examples of how the application of the theoretical frameworks discussed in Chapter 2, particularly those of Masters and Corning, can contribute to the field. Comparative politics has always been concerned with the comparability of structures, processes, and behaviors within the context of contemporary nation states. Comparativists have also been concerned with the historical development of the above and the unique contexts in which countries have developed (see Apter, 1996; Laitin, 2002). Both comparative and international politics also invite further consideration of the issues surrounding the levels of analysis problem in research, which was discussed at some length in Chapter 3. Moreover, both fields also incorporate the study of elite political behavior.

As the following discussion will, hopefully, clarify the evolution of politics, political institutions and political behavior are so interconnected that from the vantage point of biopolitics, it seems inappropriate to disconnect them by using a particular set of field or subfield categories for the discipline in order to organize our discussion. Fields and subfields coalesce around the more inclusive process of evolution. Moreover, from the perspective of biopolitics, many of the distinctions drawn between comparative and international politics are simply not relevant (see Mair, 1996). The evolution of political institutions and processes as a result of adaptation to different and changing environments over time involves both domestic and foreign policies and, hence, blurs the distinction between comparative and international politics. Therefore, the two fields merge intellectually as one considers the evolution of politics in all its manifestations.

This is even more apparent when one examines the study of political development and modernization or considers the approaches of scholars like Wallerstein (1991) whose world systems approach is consistent with an evolutionary model. A good example of how the artificial distinction of subfields is inadequate is to be found in the cases of "failed states" over time and in specific contexts of international relations. Fukuyama (2004, 2011) makes a strong case for the failure of various key institutions – both political and economic – under different conditions as a major factor in the failure of states over the course of human history. An annual survey of failed states by the journal *Foreign Policy* shows the impact of a number of specific conditions on the ability of a state to function in the international political system (see http://www.foreignpolicy.com/failed_states_index_2012_interactive). The particular case of failed democratic political systems has been carefully reviewed by Diamond (1999, 2008) and Svolik (2008). Both have focused on the concept of consolidated democracies – those that have sustained political order over an extended period of time and are most resistant to lapses into authoritarian rule. Those that remain vulnerable are termed transitional by Svolik (2008, pp. 153–154) and are the subject of his empirical analysis wherein he finds that "the level of economic development, type of democratic executive, and type of authoritarian past are key variables in determining the likelihood of consolidation" (Svolik, 2008, p. 153).

Our own view of political evolution (Corning & Hines, 1988) suggests that one line of potentially fruitful research would be to study and characterize innovative adaptations (political evolution) and their spread or demise over time. These political innovations that mark an evolutionary change often have taken the form of paradigms of political order (e.g., city states, ancient empires, the modern nation-state, modern empires, confederations, federal systems, and contemporary experiments like the European

Union) and discrete institutions (e.g., modern bureaucracies, military institutions, legislatures, political parties, and interest groups) that, due to their success in certain settings, have been replicated, although not always with subsequent success, in other environments. Wallerstein's (1991) work looks at the evolution of world systems in a similar way and the two research programs are quite compatible and share much in common in their theoretical underpinnings. For these reasons, we advocate the use of the terms world politics and international relations as being more reflective of the contextuality and configurative character of politics at all levels, both in the past and in the contemporary world. Kahler has observed that "Both international relations and comparative politics have spent much less time considering the political units within which identity and order are constituted" (2002, p. 69). We would add that both fields have largely ignored the evolutionary factors related to ethnicity, ethnic conflict, and intragroup conflict that contribute significantly over time to whether order or disorder – Huntington's "political decay" (1968) – will occur.

In what follows, we will review several lines of inquiry that have been pursued by proponents of biopolitics in both comparative and international politics. First, we will look at comparative politics and comparative political analysis and argue the need for greater contextuality and range of comparison throughout all of human existence. Then, we will examine some early works on international politics within the biopolitical frameworks of Pettman (1975, 1981) and Wiegele (1979a) as well as Falger's (1994, 1997) interpretations of the field of international politics and Thayer's (2004) key study of war and ethnic conflict and therein the applicability of evolutionary and ethological studies of peace and war and the relevance of examinations of the "politics of identity" (ethnicity, xenophobia, and patriotism). The discussion of the evolutionary origins of politics and the state, although discussed in the section on comparative politics, is equally relevant to international politics. Indeed, one might argue that the political evolution of systems of politics-world systems is as applicable to the study of international politics as it is to comparative politics.

COMPARATIVE POLITICS AND COMPARATIVE POLITICAL ANALYSIS

The field of comparative politics, despite the challenges presented by Somit and Peterson (2005) in their critique of the prevailing views on democratic nation building, continues to focus basically on the same themes and research areas that have characterized the field since the post-World War II (WWII) study of modernization and development spurred a large body of work that was undertaken in support of the goal of extending the scope of democratic political systems in the aftermath of the defeat of authoritarian regimes in Germany, Italy, and Japan (Apter, 1996; Chilcote, 1994; Laitin, 2002; Mair, 1996). Most recently, Laitin provided an organizational framework for comparative politics that reflects the growing emphasis on institutionalism, the state, and the problem of order, in particular. Laitin (2002) sees an emergent subdiscipline of comparative politics that is characterized by a tripartite methodology that includes statistics, formalization, and narrative. He then explores the three major "consequential political outcomes" (at p. 633) – democracy, order, and forms of capitalism. Although he provides no references to the literature of biopolitics, these three outcomes equate to key areas identified in biopolitics:

- Democracy: The relationship of human nature (dominance, hierarchy, and ethnicity) to the prospects for democracy and world politics (Somit & Peterson, 2005; Vanhanen, 2003, 2009, 2013).
- Order: The evolutionary/adaptationist strategies leading to the creation and sustainability of social/political order (Corning & Hines, 1988; Diamond, 1997, 2008; Fukuyama, 2002, 2004, 2011; Hines, 1983; Johnson, 1995; Masters, 1989; Sanderson, 2001, 2014; Thayer, 2004).
- Forms of capitalism: The evolution of bioeconomics and the organized complexity explicated in the works of Corning (2003, 2005) as well as the challenges brought by the expansion of a capitalist world economy and globalism (Thayer, 2004; Wallerstein, 1974, 1979, 1980, 1984, 1989, 1991).

Vanhanen (2013) has provided a specific empirical study examining the relationship between intelligence as expressed in national IQ data and democracy. The research conducted by Lynn and Vanhannen (2002, 2006, 2012) demonstrated that differences in IQ "... explained a significant part of the global variation in economic, educational, political, health and some other measures of human conditions" (Vanhanen, 2013, p. 53). Vanhannen constructed an Index of Power Resources (IPR) that combines "... four basic evolutionary variables: students, literates, family farms, and the degree of decentralization of economic power resources ..." (2013, p. 54). Combining evolutionary and environmental explanatory variables, his analysis argues for social scientists to take the evolutionary roots of human nature and the necessity of taking evolved characteristics of human nature into account seriously rather than continuing to rely solely on environmental variables (see also Vanhanen, 2003, 2009). While this research is comparative, it also addresses the global impact of the variables used and engages a wide range of international relations issues – health, education, global economics, and power relations.

In an earlier review of the literature of biopolitics in the mainstream field of comparative politics, Watts referred to the contributions that have been made as "scattered outcroppings in the comparative terrain where life science insights applied by political scientists have attempted to make a contribution, and may have expanded those areas in useful ways" (1994, p. 209). This is probably still a fair characterization of the contributions to date in this field, at least in so far as the contributions are specifically efforts to address comparative politics as a subfield of political science.

Perhaps the greatest contribution of biopolitics to comparative politics has been to enlarge the field dramatically. For example, by advocating the comparison of nonhuman primate behavior to human behavior and by recognizing the importance of human prehistory and early history, proponents of biopolitics have reconfigured comparative analysis to include a much wider range of subjects of study than has constituted the traditional subfield of comparative politics. Another important point, not taken into consideration by Watts, is the fact that the most dynamic movement in comparative politics, the study of political development and modernization, was replete with works (e.g., Almond, 1990) that were informed by structural functionalism and were at least implicitly evolutionary in as much as they sought to explain social change in the West and then in the non-western world as developmental. This period of significant work during the 1960s and 1970s (which includes the first of Wallerstein's major works) also saw the competition between comparative political analysis and area studies or regional studies. The former sought generalizations, including probabilistic explanations using aggregate data analysis, across space and time (aspiring to nomothetic explanations), whereas the latter emphasized the idiographic interpretation of distinctive developments in specific countries and regions (historical and contextual explanations).

This intellectual tension is reconceptualized within an evolutionary framework as the transfer of institutions and political processes that represent evolutionary adaptations and survival strategies in a particular setting to other settings (ecological niches) where they may or may not prove successful. We discuss this matter further in Chapter 7. As an examination of work on the origins of the state demonstrates, the political evolution of the state represents the first major example of a specific adaptation that has been successfully transferred to other environments over time, albeit in a variety of forms as might be expected given the qualitative features of various settings. Thus, the historical and configurative analysis of the area specialist complements the attempts at generalization through historical and developmental analysis. Political evolution is reserved to define those innovative adaptations that resulted in new structures and processes being created to perform political functions.

As early as 1968, Harold Lasswell, writing in the inaugural issue of the journal *Comparative Politics*, argued for a comparative method that was inclusive and that adhered to the "requirement of contextuality." He observed that the comparative method as then employed by political scientists was "insufficiently contextual, inadequately problem-oriented, and unnecessarily restrictive in technique. In brief, it has been insufficiently configurative" (1968, p. 6). He went on argue for the "recognition of the role of the political process in prehistoric as well as in historic times" (Lasswell, 1968, p. 6). He went so far as to recommend the "formidable expansion of research on nonhuman societies" as an opportunity to compare and contrast genetic constitutions in relation to variable environmental contexts. Lasswell's "contextuality" is a useful condensation of the argument made in Chapter 3 about the importance of using multiple research strategies and methods at different levels of analysis to gain greater understanding of the contextuality of events. This is a straightforward call for the application of the interactionist paradigm and of evolutionary theory to the study of political change even though the vocabulary of Lasswell was not explicitly derived from evolutionary theory. Lasswell's emphasis on the dimension of time and on developmental analysis underscores, as did Thorson (1970), the significance of human evolution in the larger context of biological evolution.

It is also worth noting that the work of Wallerstein (1991) with its emphasis on the historic development of world systems that constitute successful macro-level adaptations fits this configurative analysis program of Lasswell. Moreover, Wallerstein (1991) strongly advocates changing the paradigm of social science to a historical, developmental paradigm, which, although not explicitly evolutionary, nonetheless is entirely consistent with a general systems and evolutionary paradigm (see Wallerstein, 1991 in particular).

In a recent review of the field of comparative politics in *The New Handbook of Political Science*, Mair claims that while the tendency of comparative political research in the 1950s and 1960s was to emphasize "universal relationships, and thus global comparisons, the tendency within comparative research over the past decade or so has been to move away from general theory by emphasizing the relevance of context" (1996, p. 328). Reflecting much the same tone and sensitivity to the tension between the general and the particular, Apter (1996) compares the old institutional approach (before WWII), the new comparative politics (post-WWII through the 1960s), and the neo-institutionalism that he sees currently leading the field in popularity. This most recent trend is very consistent with biopolitics' emphasis on political evolution as adaptation to context (environment, ecological niche) and on the importance of determining what caused the success or failure of that adaptation over time.

As the debates over the claims for political development and modernization raged in the 1960s and 1970s, there was at least one political scientist, Braibanti (1969, 1976), whose background was in comparative administration and bureaucracies and who was influenced by the work of Lasswell and the importance of contextuality (Hines, 1978). His treatment of political development displays a concern for the tension between external (usually Western) and indigenous values in Third World or developing countries. The political challenge of changing culture and institutions in pursuit of political modernization and development was enormous, and frankly, was seldom met. Even today we see how difficult this process is throughout the world.

Braibanti (1969) saw the process of political development as including four essential elements: architectonics, diffusion of power, institutions, and innovations. Although he never used the term political evolution, these four elements reflect the essence of what that term implies, namely, the successful adaptation in the first instance or subsequently as the innovation diffuses of a new strategy for collective survival that involves new elements of culture, institutions, and participation. Braibanti offered the concept of architectonics to describe the complex process of the adaptation of novel political forms (e.g., constitutions, bureaucracies) into indigenous societies where the most fundamental change that is required is in the realization of "common agreement on a fundamental polity of the state – an overarching purpose which gives form, cohesion, and direction to all public action within a sensed community" (1969, p. 37). This is another way of describing the process that Masters (1989) describes (see Chapter 2 and the discussion below), whereby the natural tendency toward behaviors that would be explained by inclusive fitness can give way to reciprocity relationships that Masters refers to as mutual aid and sociability and that Corning (1983) would identify as examples of synergy.

Consistent with Masters and Corning, Somit and Peterson's discussion of the human capacity for "indoctrination," which is also described in Chapter 2, provides the biopolitical framework into which the elements of Braibanti's concept of political-administrative development can be seated. The same compatibility exists with the series on political development sponsored by the Social Science Research Council, especially the volume on crises and sequences of political development (Binder, 1971; see also Almond, 1990). The biopolitical framework helps identify the essential research problem (achieving political and social order in the face of a human behavioral repertoire that is somewhat limiting for altruism and sociability) and the concrete, historical analysis of specific cases of political development along the dimensions of architectonics, institutions (including rule of law, Braibanti, 1969), diffusion of power through social participation (see Masters, 1989), and innovation, of which the origin of the state is the quintessential example.

We are convinced that comparative political research thus becomes much more inclusive and more susceptible to explanation using both proximate and ultimate types of explanations within the evolutionary framework of biopolitics. Whether one is interested, as Roger Masters (1989) and Masters and Way (1996) have been, in comparing individual reactions to political candidates in different countries using ethological methods, or in comparing the effects of age and age structure on collective political decision making at the level of local governments using a variety of methods of data collection and analysis (Schubert, Wiegele, & Hines, 1987), or on the evolution of world systems (Wallerstein, 1991), the overall framework remains one of comparing the effects of variable contexts (e.g., structures and cultures) on recognizable patterns of human behavior (e.g., hierarchies and dominance structures) that are part of our evolved behavioral repertoire as a species. As Sullivan and Masters describe the framework for their own research: "Humans, like other primate species, have evolved an elaborate system of expressive displays that function in relations of dominance as well as in other social interactions. The concepts and methods derived from ethology and social psychology allow us to explore the way this system functions as a means of integrating social information and cultural expectations in popular responses to political leaders" (Sullivan & Masters, 1994, p. 237).

Over the past 6 years, Fowler and Christakis (2008) and Hatemi and collaborators have produced a body of empirical studies. Fowler and collaborators have explored the impact of genetics on political participation and voter turnout. Hatemi and colleagues (Hatemi et al., 2007, 2009) have produced a substantial body of empirical research consistent with the evolutionary model and the biopolitical theory. Their findings are discussed in Hatemi and McDermott's edited volume, *Man Is by Nature a Political Animal: Evolution, Biology, and Politics* (2011). The publications of this team of scholars on the genetic, physiological, and endocrinological sources of individual differences in political attitudes, fear dispositions, mate selection, addiction, personality, political violence, and religion provide a model for scholars of politics and the life sciences. Their research also illustrates how an interdisciplinary research approach could be applied in comparative politics to explore the complex relationship among genetics, phenotypic behavior, and politics and, thereby, reveals how different environments and cultures interplay with genetic inheritance.

In pursuing this kind of comparative research, we concur with Watts when he urges that we must avoid naive reductionism and "conceptualize human behavioral and organizational plasticity as a series of 'potentials' that are differentially augmented and rewarded depending on environmental contingencies. Biological organisms interact with their environments in a mutually-interactive process of epigenesis in which both biological potential and environmental conditions play a role in the final form sociopolitical behavior takes" (1994, p. 233). The emergence of specific institutions at specific times and in specific places take the form of successful adaptations to an environmental setting (political evolution) is exactly the phenomenon to which Watts refers. The focus within comparative politics that has the most immediate affinity with a biopolitical approach is the study of political development and modernization. We shall now turn to an examination of that literature.

POLITICAL EVOLUTION: THE CASE OF THE ORIGINS OF THE STATE

As has already been suggested, one of the most robust and contentious areas of research in political science during the past few decades has been the study of the processes and outcomes of what has been generally labeled "political development" and "modernization" (see Almond, 1990; Apter, 1996; Chilcote, 1994; Mair, 1996; Rustow & Erickson, 1991; Wiarda, 1985). Among proponents of biopolitics, Corning and Hines (1988), after reviewing debates in this literature, have argued for the incorporation of these key concepts of political development and modernization into a larger process, the process of political evolution. Their approach is intended to place the study of these concepts into, historically speaking, a much longer and more encompassing frame of reference that extends from the earliest human settlements to the present global system. This approach is largely in sympathy with the approach of Wallerstein, although the focus is more on the discrete cases of political evolution at the societal level rather than on the macro-level of the world system.

A particular example of a research question that invites the use of the concept of political evolution as constructed by Corning and Hines (1988) is the question of when, where, and why the state emerges. Political scientists have generally ignored the question of the origins of the state (Hines, 1983) and have not recognized the possibility that there is a biological (evolutionary) basis for politics (Johnson, 1995; Masters, 1989). Indeed, outside of biopolitics scholars, the work of political anthropologists in general has been sadly ignored by the discipline of political science, even though attention was directed toward the potential contributions of anthropology by Friedrich as early as 1968. As indicated above in the discussion of political evolution, the emergence of critically important new political institutions and the detailed description of the circumstances (environmental factors) associated with their emergence is a major focal point of comparative political inquiry from a biopolitical perspective. Political anthropologists have been keenly interested in these historical developments and there is a considerable literature that addresses the question of the origins and elaboration of the state, its precursors, its spread, and its component institutional parts (see Corning, 1983; Fukuyama, 2011; Geiger, 1985, 1988; Hines, 1983; Johnson, 1995; Masters, 1989, for citations of this literature).

The state, like other institutions, emerges in response to historical and ecological conditions that serve as a stimulus for human beings to create and pursue alternative strategies for survival and for the attainment of other values and goals as reflected in human culture. Although there are many different interpretations and explanations offered for the origin of the state, we cannot review them in detail here. Some observers emphasize conflict and balance of power dynamics, others emergence through innovative behaviors of leaders and the manipulation of symbols, others feedback mechanisms and complexity, and still others, ecological limitations and population pressures.

After reviewing all these alternatives from the perspective of evolutionary biology, Masters' notes that the emergence of the state is especially hard to explain because "[s]ocial cooperation in very large groups of animals seems to violate the basic premises of the neo-Darwinian theory of natural selection ....[I]f natural selection primarily tends to favor the reproductive success of individuals, cooperation or self-sacrifice that benefits unrelated strangers ... should be less adaptive than selfish or nepotistic behavior benefiting close kin and reciprocating conspecifics" (1989, p. 153). After reviewing the implications of the inclusive fitness theory and the problem of "free riders" and the notion of reciprocation among unrelated individuals (reciprocal altruism), he proceeds to show how recent studies of the possibility of group selection lead to the conclusion that: "since contemporary theories of natural selection stress the priority of individual reproductive success, one can never presume that a behavior has been selected because it redounds to the benefit of the population or species. But it is equally erroneous to assume that ... natural selection never generates behavior of this sort" (1989, p. 161). He posits a dialectical relationship between individual inclusive fitness and group interest, particularly in a complex social species like *Homo sapiens*. This is particularly the case for humans whose larger brains and complex linguistic systems have allowed them to embrace a much wider range of social behaviors than other animals. This is exactly the dynamic and dialectical process that is described by those who have sought to explain how the state emerges in human history from a biological and anthropological perspective (see Alexander, 1979, 1987, and numerous references in Hines, 1983; Johnson, 1995).

INTERNATIONAL RELATIONS, WORLD POLITICS, AND BIOPOLITICS

The expressions international relations and world politics are used interchangeably here, rather than international politics or studies, because we believe it is more consistent with the biopolitical perspective which, following ecology, stresses the interrelatedness of phenomena at various levels and because it reflects the conflation of comparative and international politics. As Falger observes, "the relationship between biopolitics and international relations is not self-evident. Leading biopoliticians do not often think of international relations when they introduce the biological approach to their readers" (1994, p. 117). Wiegele, himself a leading proponent of biopolitics and specialist in international relations, also commented that "Perhaps no association seems as remote as that between biology and international relations. It is usually assumed that the 'high politics' that so often characterises relations between nations assures the isolation of international affairs from the more mundane aspects of political life" (1979, p. 101). Despite these assessments of the lack of receptivity to biopolitics in the study of international relations, there is actually a substantial and growing body of scholarship that explores elite behavior in decision making, the relationship of the human behavioral repertoire to international conflict, and the biopolicy of environmental and health issues (see Chapters 6 and 7 of this volume for a discussion and sources).

The first appearance of a challenge to take evolution and the life sciences seriously in the study of international relations, leaving aside more historical forebears (see Falger, 1994, 1997; Wiegele, 1979a), may well have been the work of Sprouts (1965, 1968) in introducing ecology and an article by Masters (1964) comparing world politics to a primitive political system, thus introducing both the ethology and the anthropology of the early state to readers. Following closely was Corning's (1971) first attempt to present a more comprehensive and avowedly evolutionary view of the relation of the life sciences to the study of politics, including international relations.

Wiegele credits Mills (1973) and Pettman (1975) as being trendsetters in the 1970s. Mills (1973) argued that policy issues, including food supply, reproduction, and human adaptability that are of concern to and addressed by the life sciences, have become important features of modern international relations. Pettman went so far as to claim that:

> The growth of inter-disciplinary interest has substantially enhanced the number of levels that can be related to a political explanation for world affairs, with the horizontal expansion into other disciplines further encouraging the number of vertical levels brought to bear. The contemporary study of world politics now ranges from the evolutionary ascent of humankind to his possible ecological demise, from Betelgeuse to the man in the street, from the machinery and social performance of the brain to the group behaviour of baboons. It moves with more or less ease from biology to cosmology, from psychology to physiology, and from ethology to anthropology, and on the vertical plane from the gene through multiform man to his world systems and societies. (1975, p. 24)

Pettman is quite adamant about the need for students of international relations to press on beyond the mainstream topics of their research and explore the vertical and horizontal dimensions noted above. How, for example, can we talk about war, a central question in the study of world politics without talking about conflict in general of which war is a particular, albeit especially violent, form? How can we talk about conflict in general and war in particular without talking about aggression or self-transcendence, loyalty, obedience, or the urge to survive? And how, in all humility, can we pronounce on drives, needs, and capacities like these without some associated knowledge of psychology, biology, ethology, and brain physiology? To do less would be to abuse the real scope of the problem in the interests either of analytical precision, misplaced academic modesty, activist fashion, or some such defense of a provincial status quo. Likewise, how can we talk about imperialism and the profit mechanics for the market economy without discussing dishonesty and greed? And these in turn may well have to be seen as human attributes of a biological or cultural-anthropological kind, or as psychological or sociological processes, before their full political implications become clear (Pettman, 1975, p. 27).

Pettman provides a lengthy discussion of many of these vertical and horizontal linkages and sees real prospects for the expansion of the study of international relations to incorporate insights from the life sciences. Despite Pettman's enthusiasm, the impact of biopolitics and the life sciences on the teaching of international relations has been limited until recently. Somit, Seo, and Peterson (1994) in reviewing textbooks in international relations found that there was a only a modest, positive trend in the number of references in textbooks to the relevance of ethological and sociobiological concepts, particularly as applicable to the study of violence, war, and conflict. However, they also observed that these references often reveal that the authors are still relying on the outdated and popularized work of Lorenz and Ardrey rather than on contemporary research and that they seem to be unaware of the work of their fellow political scientists working within a biopolitics paradigm (Somit, Seo, & Peterson, 1994, p. 138).

As was the case in comparative politics, we find the early work of Lasswell, beginning in the 1940s, to have prefigured much of what today constitutes a biopolitical approach to international relations. Lasswell's "developmental analysis" and his articulation of the "world manifold of events" represent an early attempt to show the seamlessness of the vertical and horizontal dimensions referred to by Pettman (see Hines, 1998 for a review of Lasswell's contributions and citations; Eulau, 1963; Lasswell, 1970). Today, the work of Wallerstein (1974, 1979, 1980, 1984, 1991) presents somewhat the same challenge to the social sciences to produce work that is more contextual and configurative in its analysis. As he puts it, "everything that is historic is systemic and everything that is systemic is historic" (Wallerstein, 1991, p. 229).

Wiegele (1979a), in reviewing the biology of international relations, organizes his discussion into research on the nation, the international system (although without mention of Wallerstein's work), and international crises (with emphasis on the behavior of elites under stress during international crises) and then deals with the biopolitics of political elites and human conflict separately. Falger structures his discussion around three key questions in international relations: "(1) the causes of war and the conditions of peace/security/order; an essential subsidiary problem is the nature of international power; (2) the essential actors and/or units of analysis; (3) images of the world/system/society of states" (1994, p. 116). He then interprets international relations research in light of the framework of biopolitics. Both of these scholars conclude that there is much that biopolitics has contributed and can contribute to the study of international relations.

Falger (1994, p. 119) reminds us that the realist strain in international relations theory that emphasizes the study of power and that both realists and Marxists recognize that human beings and nation states are self-interested and tend to behave accordingly, hence the idea of national self-interest. He cites the claim of Waltz that "[I]nternational politics like all politics is a struggle for power. Whatever the ultimate aims of international politics, power is always the immediate aim" (1979, p. 117). This quest for power is then understandable in the context of the ultimate evolutionary goal of survival for the individuals and the structures (e.g., the state) that are engaged in the pursuit of power.

Falger (1994, p. 119) also points out that at least two well-known figures in international relations, Morgenthau (1967) and Wright (1955), have been influenced by biology (see also Thayer, 2004). The former understands that "[t]he drives to live, to propagate and to dominate are common to all men" (Morgenthau, 1967, p. 31; see also Falger, 1997). Wright (1965) recognizes the level-of-analysis problem and argues for a research program that moves from the level of states and societies and the level of individual decision makers always to be understood "in relation to the international environment in which the actions take place" (Falger, 1994, p. 121). Falger claims that "it would be easy for biopolitical scientists to add relevant information and methodology at the individual level" and that the study of "group formation and, in particular, the origin of the state itself are highly relevant for the field of international relations, although so far underutilized" (1994, p. 121). He also urges that more attention be given to the differentiation of ultimate and proximate causes, noting that Waltz raises the question of whether the causes of war are best understood in terms of "human aggression on the level of individuals, in the nature of organized states or societies, or in the nature of the international system itself" (Falger, 1994, p. 121). Waltz's own conclusion that the context of anarchy in the international system is a necessary precondition for war invites the further consideration of the individual decision makers and the conditions existing in the particular states and societies involved in conflict.

Although not mentioned specifically, we would add (consistent with the levels of reality and analysis in Masters' model) that the rhetoric and symbols used by the decision makers are relevant and must be included in an analysis that appreciates the importance of the role of culture in explaining the occurrence of war. Lasswell (1970), particularly in his work on propaganda, understood this dimension all too well as does Johnson (1986, 1987) as shown in his work on the use of pseudo-kinship terms (fatherland, motherland). Eibl-Eibesfeldt (1979), one of the foremost figures in ethology, has put this in context in his study of the biology of peace and war. Our capacity to define others out of humanity-cultural pseudo speciation in Eibl-Eibesfeldt's framework makes it possible to engage in conflict with conspecifics by shifting the conflict to the interspecific level.

> Over the biological norm filter that inhibits destructive aggression in man as in other creatures, a cultural nom filter is superimposed that commands us to kill. The important point to bear in mind is that destructive war is the result of cultural evolution. Furthermore, it is not, as is sometimes maintained, a pathological phenomenon, but performs important functions .... It also accelerated biological and cultural evolution by intensification of selective pressures. This applied both to the rapid development of the brain and to the development of altruistic behavior. The question remains open whether humanity can break out of this self-reinforcing process of increasing aggressivity or is bound to go on passively subjecting itself to it. (Eibl-Eibesfeldt, 1979, p. 123)

This capacity helps explain why we can overcome the limitations of inclusive fitness strategies to embrace reciprocity and strategies of mutual aid and sociability despite our natural tendencies to behave in an entirely self-interested way. It is this capacity to extend the in-group beyond kin to include fellow countrymen that, when joined with our natural tendencies to form dominance structures and hierarchies, enables us to construct the state (see Corning, 1983, p. 310 ff; Masters, 1989; Shaw & Wong, 1989; White, 1997, for more detailed discussions of this topic).

According to Wilson (2012), kin selection has given way to group selection because the latter provides more explanatory power over the course of cultural evolution giving rise to what Wilson has defined as cooperative social and cultural structures that are the outcomes of "eusocial" evolution. "Multilevel selection consists of the interaction between forces of selection that target traits of individual members and other forces of selection that target traits of the group as a whole. The new theory is meant to replace the traditional theory based on pedigree kinship or some comparable measure of genetic relatedness" (Wilson, 2012, pp. 54–55). He goes on to say, "[T]he origin of eusociality, in which organisms behave in the opposite manner, has been rare in the history of life because group selection must be exceptionally powerful to relax the grip of individual selection. Only then can it modify the conservative effect of individual selection and introduce highly cooperative behavior into the physiology and behavior of the group members" (Wilson, 2012, p. 55). Thus, there is an inherent tension, prone to lead to conflict, between these two processes: individual and group selection. The tension and conflict are then played out in discrete ecological niches by individuals within groups and by groups in a larger social–cultural–political context. Therefore, Wilson (2012, p. 57) concludes: "To form groups, drawing visceral comfort and pride from familiar fellowship, and to defend the group enthusiastically against rival groups – these are among the absolute universals of human nature and hence of culture."

This complex, multilevel drama plays out in politics on a variety of stages with the results sometimes leading to conflict (e.g., war) and sometimes leading to cooperation (e.g., associations, alliances, coalitions, and contracts). The role of political elites and their decision making is critical to the determination of which will prevail at any particular time. Returning to Wiegele's research, his review includes an extensive discussion of his own research (Wiegele, 1973, 1976, 1977a, 1977b, 1978, 1979b, 1982) on elite decision making under stress and Falger cites the same works in his discussion of the relevance of biopolitical methods for the study of individual and small group behavior in decision making. Clearly, the rational actor and organizational behavior and process models of Allison need to be complemented by this kind of rigorous empirical research however difficult it may be and however much it may rely on remote assessment and interpretation of recorded materials (see Salter, 1996).

Falger (1994) briefly discusses man–environment relationships in his review, but cites only Pirages (1978). We have treated this dimension extensively in Chapter 7 in the context of ecological politics and the biopolicy of the environment and will not deal with it further here except to note that the relevant literature is far more extensive although often not explicitly biopolitical. McDermott (2004, 2009) has dealt with elite decision making under stress in her study on presidential leadership and illness (2007) and risk taking and political psychology in international relations, all in the context of an evolutionary framework (Hatemi & McDermott, 2011; Lopez & McDermott, 2012).

Probably the most fruitful line of research in international relations for biopolitics has been the growing body of scholarship devoted to the evolutionary origins of ethnicity, ethnic conflict and war, and the potential for achieving cooperation by creating a framework for political evolution that achieves sustainable order in societies and presents the possibility of attaining a stable global community where the likelihood of major wars is minimized. Begining with the hypothesized Hobbesian state of nature, political scientists working in IR have often characterized the state of world politics as anarchic with the quest for world order never ending. Theorists of realism, idealism, liberalism, and rational choice have argued their cases for howbest to achieve world order and have demonstrated how the historical paradigms of political order (city-state, ancient empire, nation-state, modern empire, balance of power systems and contemporary experiments like the European Union) have failed to provide that hope for stability. The hypothesized "end of history" (Fukuyama, 2004; see also Mueller, 2014) has not been realized as new challenges to the spread of liberal democracy have emerged. The failed Arab Spring, resurgent Russian nationalism, Putin's seizure of Crimea and the apparent threat to the independence of the Ukraine, China's growing assertion of its power, global terror, the recent global recession, competition for limited resources (e.g., oil, water, rare minerals), demographic pressures, and the omnipresence of ethnic conflict, as well as the overthrow of democratic political systems in Nigeria, Thailand, Venezuela, Bangladesh, and others (Diamond, 2008, p. 36) all point to the prospect of crises arising in world politics and international relations. Fukuyama's (2011) recent focus on the problem of political order in historical and current perspective is indicative of our current dilemma.

Peterson (2013) has recently examined the anarchist views of individualists and communists and has shown how cooperation has evolved from the evolutionary bedrock of kin selection, reciprocity, and group selection. Other researchers (Falger, 1991, 1994, 2001a, 2001b; Thayer, 2004; Van der Dennen, 1995, 1999) have also shown how conflict and cooperation at varying levels are, in essence, both sides of the same coin from the standpoint of evolutionary origins.

The recent work of Bradley Thayer (2004) offers perhaps the most comprehensive review of the evolutionary foundations of war and conflict as those phenomena relate to the dynamics of international relations. Thayer provides an excellent, comprehensive review of evolutionary theory and its applicability to the study of international relations arguing that our understanding of ethnic conflict and war will always be insufficient if it ignores the findings of evolutionary biology about the human behavioral repertoire. Proximate causal explanations are insufficient. In addition, we must include the ultimate causal explanatory power of the evolution of our species. He reviews the dominant theories in international relations – realism and rational choice theory. Furthermore, he rejects the two underpinning ideas that have been part of realism – the idea that man is inherently evil and the view that humans possess a drive to dominate (Thayer, 2004, chapter 2). The lack of world government, which is not the same as anarchy, creates an environment that support "offensive realism." For the rational choice theorist, evolutionary psychology provides an explanation for how humans have developed a rational mind capable of calculation, detection of cheating, and capable of deception, all of which supports the major tenets of the game theory. Thayer (2004, p. 86) cites Boyd and Richardson who state that evolutionary theory "predicts that natural selection should produce *Homo economicus*" (1980, p. 97). Meticulously researched, Thayer's analysis and arguments are buttressed by extensive research from evolutionary theory and evolutionary psychology.

Thayer shows how Darwinian natural selection, inclusive of fitness and group selection, explains that ethnic conflict and warfare can contribute to human survival and success. Warfare is fundamental to defending and attaining resources. An evolutionary version of realism and the rational choice theory shows why cooperation among unrelated groups is difficult to achieve and why ethnic identity remains such a powerful force in politics at all levels. In addition, Corning (2003, 2005), through his synergism hypothesis and explication of organized complexity, and Fukuyama (1999, 2011), in his extensive examination of the evolution of political order, cover these topics thoroughly as well. All three provide detailed cases to support their arguments.

THE CASE OF ETHNIC CONFLICT AND RECURRING WARS

Many international conflicts are recurrent, and some of these are characterized by periods of violence, including war, that are hard to describe as planned products of rational decision making. Such conflicts account for 40–50 percent of all state-against-state wars and are arguably the predominant form of warfare today. Similarly, internal civil wars, fueled by deep enduring animosities, continue to spill over their borders, often embroiling, directly or indirectly, major powers in system-defining confrontations. Not surprisingly, there have been many theories as to what mechanisms sustain a conflict and predispose to periodic violence. According to Long and Brecke (2003), however, none of the prevailing structural and behavioral models of such conflicts suitably includes the human emotive dimension.

According to a rationalist approach to decision making, for example, such wars are the product of cost–benefit calculations, that is, if the benefit achieved from attacking outweighs the cost, an aggressive strategy is more likely to be taken. But as noted by Friend and Thayer (2011), it would seem more rational and cost-effective to avoid conflict and make mutually beneficial settlements than to wage war. The fact that this is not the case suggests that additional causal mechanisms are responsible for coalitional aggression. To the extent it is noted at all, emotion is either dismissed as a pathologically irrational factor or viewed as a potentially important variable that is inestimable. For instance, deMesquita argues that "the selection of war or peace is a choice that is initiated by individual leaders who must accept responsibility for their decisions .... Their choices depend on their estimation of costs and benefits" (1981, p. 5). Thus, deciding whether to go to war or not is a rational choice for decision makers under certain conditions, while emotion is treated as irrational and, therefore, exogenous to decision making and action. This clearly contradicts many findings of neuroscience summarized in Chapter 5.

In an effort to integrate emotion and reasoning to explain the causes of recurrent conflict, Long and Brecke (2003) present three emotive models of recurrent conflict. The first model is "frustration–anger–aggression" and is derived from the observation that humans often become aggressive when they are unable to reach a desired goal. In essence, this model argues that the thwarting of an objective interest, in particular an interest in territory, will lead to an instrumental form of aggression designed to reassert a claim or goal, for self-enhancement or self-defense. Although it does not fit the dominant rationalistic framework, the frustration–aggression hypothesis also can be viewed as less instrumental and more emotional in its motivation.

Their second model, "unacknowledged shame and rage," asserts that certain events trigger an emotionally driven set of decisions and actions that perpetuate war. Specifically, unexpected military defeat, loss of territory, or hostile occupation of territory, followed by the psychological absorption of such a calamity by a collective body, can trigger a shame–rage–aggression sequence. The result is an almost unquenchable thirst for vengeance, the compelling desire to get even, right a wrong, or avenge an injury (Harkavy, 2000). Conflict begins after one or both parties suffer a "narcissistic" injury to its (their) group identity, such as the loss of territory, and collective shame is evoked. Shame is a very powerful emotion that involves a negative evaluation of the global self, and anger provides temporary relief by mobilizing the self and avoiding feelings of condemnation. When shame goes unacknowledged, however, another person or group becomes identified with the hurt, thereby leading to further withdrawal, recrimination, and threats. Rage and vengeful violence are the ultimate forms of protection from shame. Furthermore, the shame–anger–aggression emotional sequence is recursive, as one subsequently feels ashamed of one's anger, which generates resentment and hatred because the shame driving the loop is outside awareness. Nations locked in this spiral have great difficulty deescalating or resolving conflicts of interest not because of the content of the issue but because of the emotional milieu in which conflicts of interests occur. In such instances, violence can become compulsive (Long & Brecke, 2003).

Long and Brecke's third model is "fear and perceived threat." As demonstrated in Chapter 5, fear is a very powerful survival emotion designed to facilitate or inhibit certain behavioral patterns, often without conscious awareness on the part of an individual or a collective. Whether powerful or weak, fear functions as a signal indicating the possibility of danger arising from power relations in which the party is involved. Fear has environmental triggers, some preset during evolutionary adaptation, but many others learned. The change in the environment that triggers fear need not be a hostile act or threat per se but only a novel event that one has learned to associate with threat. Contrary to our usual supposition, we do not fear because we are threatened, we feel threatened because we fear. Furthermore, the object of fear is not retrospective (the change in power balance); it is prospective, such as the imagined possibility of harm or injury resulting from a changing power balance. Fear forces attention on this environmental change, among innumerable other environmental changes, that stirs appraisal and inferences of threat and that informs preference formation, and action, including the possibility of aggression (Long & Brecke, 2003, p. 30).

Although they vary in focus, each of these models incorporates the human emotional dimension and challenge prevailing theories in political science. They also recognize the evolutionary bases of human behavior and the necessity of more interactive, inclusive theories of ethnic conflict and recurrent wars grounded in cognitive neuroscience research presented in Chapter 5 regarding decision making and group behavior.

CONCLUSIONS

As the discussion above demonstrates, the literature of biopolitics that is applicable to the study of comparative politics, world politics, and international relations is substantial in its contributions and has grown quite extensive. The evolution of political institutions, especially the emergence and institutionalization of the state, is a research topic that is centrally relevant to those fields. As we learn more about how political institutions have emerged to create and sustain political order, we will understand better how human evolution has equipped the human animal to adapt and survive. The importance of our understanding of human nature and the role it plays in defining what is quite possible and what is difficult in terms of political behavior offers much to students of politics who recognize that the achievement of social order is highly problematic. In the 21st century, the challenges presented to humankind to achieve sustainability in the face of demographic and scarce resource pressures will require further institutional adaptations to minimize the likelihood that ethnic conflict and warfare will increase as economic and military power are resorted to more often than "soft power." This dilemma has special bearing, as was illustrated in Chapter 2, with reference to Somit and Peterson's (2005) research on human nature and the prospects for the growth of democratic political orders.

CHAPTER 5

BIOLOGY AND POLITICAL BEHAVIOR

Robert H. Blank

Conceivably, the one area where the evidence of biology's impact on politics is most developed, and controversial, is the study of political behavior. As noted in Chapter 1, it is also that sphere where biological evidence traditionally has been stifled because it conflicts with the predominant SSSM which holds that human behavior is shaped solely by culture and socialization. Although recognition of the biological foundations of behavior does not negate the importance of environmental factors in understanding human behavior, the history of the debate has been an acrimonious one with a tendency to coalesce into orthodox opposing camps. As a result, biological models of behavior, even when seen as heavily moderated by environmental influences, are dismissed by many in political science. For reasons discussed here, this intransigence against a more interactive model of behavior is less tenable with each announcement of findings that link biology with behavior. As noted by Wahlke, "the rapid advance in knowledge across a broad range of biobehavioral sciences threatens to make political science obsolete if it does not impart such knowledge to its students" (1991, p. 52).

As examined in Chapter 3, this shift in emphasis requires major methodological changes. Instead of relying almost exclusively on survey research and on individual self-reporting of behavior and attitudinal scales, methods are now available to observe actual behavior using research techniques

Politics and the Life Sciences: The State of the Discipline
Research in Biopolitics, Volume 12, 125–177

ISSN: 2042-9940/doi:10.1108/S2042-994020140000012004

drawn from ethology, genetics, and noninvasive neuroimaging techniques. Verbal reports long have been found to be poor reflections of actual behavior. For instance, Gruszczynski, Balzer, Jacobs, Smith, and Hibbing (2013) contend that reliance on survey self-reports of felt emotion can yield useful findings, but it fails to measure the totality of their responses. Since physiological measures rely on readings from sensors, not on self-reports of feelings, they hold the promise of tapping certain elements of response occurring outside of conscious awareness.

Political behavior is not simply a function of socialization or conditioning, thus any approach that fails to appreciate this is destined to prove inadequate. Actual political behavior can be understood only in terms of tendencies, which are a legacy of our species' evolutionary past (Somit & Peterson, 1998). As such, it is influenced by genetic and evolutionary factors as well as resulting physiological factors such as health status, nutrition, disability, stress, and overcrowding. As noted by David Easton:

> To the extent that genetic characteristics impose limits on the behavior of individuals, this may have implications for political life .... Politically relevant traits cannot be neglected as part of the total environment in which a political system operates. The fact that political science tends to do so, does not, of course, reduce either their theoretical or empirical importance. (1965, p. 72)

Although this chapter centers on contemporary research on the genetic and neurological determinants of behavior, it also examines the impact of health status, disability, and so forth on the behavior of both leaders and the public.

BIOLOGICAL BASES OF POLITICAL BEHAVIOR

Despite the supremacy of the SSSM which views political values, attitudes, behaviors, and institutions as essentially socially determined, gradually over the past 40 years a growing awareness of evolutionary biology and the biobehavioral sciences, along with methodological advances, has sparked interest in research utilizing biological models to study political behavior. In turn, the emergence of this research has spawned a paradigm shift toward the inclusion of biological influences and the acknowledgment of the codependence between genes and environment in shaping political attitudes and behaviors (Hatemi & McDermott, 2011, 2012a). Although the innate predisposition to be political was recognized by Aristotle and other early thinkers as a fundamental aspect of societal life, the idea that genes influenced behavior was summarily dismissed by the modern-day political science discipline (Charney, 2008b). The detached social interactions emblematic of modern society are considered too recent and context dependent to be shaped by evolutionary forces or influenced by biological differences. Consequently, one of the most enduring and contentious debates in Western history revolves around the relative importance of biological and environmental influences on human traits (nature vs. nurture). One presumption of the SSSM is that humans are unique compared to other species because we can and have transcended our evolution. Because of this transcendence, social learning approaches stipulate that the intergenerational transmission of political preferences can only occur through social mechanisms. Despite mounting evidence to the contrary, culture and nature are viewed as separate and opposing forces (Charney, 2008b).

A recent shift in perspective by both life and social scientists, however, that emphasizes the interplay between genes and the environment, and gene-culture coevolution, has proven more accurate than any position favoring either nature or nurture and spawned considerable research (Fowler & Schreiber, 2008). Today, a generation after Eaves and Eysenck (1974) first explored the genetic influences on political values using a classical twin study and found that parent and adult child concordance appeared to be a function of genetic transmission and personal experience rather than social learning, there is a proliferation of literature exploring genetic, neurological, physiological, and hormonal influences on political attitudes, ideologies, political participation, voter choice, political trust, political sophistication, party identification, out-groups, and political violence. Although work on the biological foundations of political behavior has lagged behind the study of behavior in general and has not been methodical, a significant literature is evolving.

As noted by Friend and Thayer (2011), since the birth of behavioralism in the 1950s, political science has focused on identifying causation anywhere but with the individual. Organizational approaches, the study of bureaucracy, and systemic analyses have practically forced studies of individuals and psychological approaches out of the discipline. However, with the advance of human genetics and neuroscience, we are witnessing the renaissance of individual approaches and a slow revival of the individual level of analysis for causation of political behavior. "Rational choice will wane, political psychology will wax as the center of the discipline. The glorification of economics and economic approaches will fade too. It is ironic that psychology was once dismissed as the 'softest' of the social sciences, while economics was triumphed as the most scientific. Yet the advance of the life sciences, evolutionary psychology, cognitive psychology and neuroscience altered this ordering" (Friend & Thayer, 2011, p. 250). Moreover, while methodological problems still plague this research, the major barriers continue to be primarily political in nature. With the paradigm shift advanced here, however, political science would be much more amenable to the inclusion of biological variables for understanding political behavior.

Ultimately, the controversy over behavioral genetics will focus on the brain, because it is through the neural system that genetic influences are manifested. Furthermore, the more we understand about the functioning of the brain, the more we are led to the conclusion that as individuals we are limited by what our brains permit. This is not to be construed as suggesting that our brains determine behavior, but rather that they mediate all genetic and environmental influences. For Roger Masters:

> The neurochemistry of behavior is not the same theory as genetic determinism. On the contrary, neurotransmitters like serotonin vary from one individual to another for many reasons, including the individual's life experience, social status and diet. Genes may influence neurochemistry. So do behavior, culture, and the social environment. (1994, p. xiv)

The brain does, however, affect or mediate every action and thought, both of political leaders and citizens. Our capacity for enjoyment, suffering, and behavior to some degree is inscribed in neurons and synapses. As a result, our interpretation of the world, including the political and social dimensions, and our responses to it are dependent on the internal organization of the brain. Therefore, in order to make sense of human behavior we must understand the organization and functioning of the system that controls or modulates it, the central nervous system (CNS). Moreover, because they serve as important communication links between neurons, neurotransmitters are a prime target for explanations of behavior. Neurotransmitters that appear most crucial to behavior are dopamine, which is crucial to the regulation of motor behavior; serotonin, which handles much of impulse transmission; norepinephrine, which is involved in transmitting impulses to the autonomic nervous system; and gamma aminobutyric acid (GABA), which inhibits neurons from firing or sending impulses. Greater understanding of these neurotransmitters (as well as epinephrine, glycine, glutamic acids, and the endorphins) is likely to provide a better appreciation of the role of the brain in human emotion and behavior.

The findings of neuroscience, then, require a model, which acknowledges that the brain has a major role in explaining behavior, that it is not just an empty organism. As observed by Pierre Changeux: "The development of the neurosciences has brought another way of looking at behaviour .... The neuronal content of the black box can no longer be ignored. On the contrary, all forms of behavior mobilize distinct sets of nerve cells, and is at their level that the final explanation of behavior must be sought" (1997, p. 97). Although it is doubtful whether we will ever be able to explain a particular behavior in terms of specific neuronal activity, it is crucial that this neural dimension be an integral part of any respectable model of behavior. Like genetics, to ignore the role of the brain is no longer possible in light of what we now know even in the rather primitive stages of neuroscience.

Current behavioral science theory, therefore, must be qualified by findings in the brain sciences as well as by genetics. Although distressing to traditional political scientists, emerging knowledge of the genes and the brain must be incorporated in order to retain credibility as a science. This was recognized over three decades ago by Elliott White when he stated: "Even now the new developments in the biochemistry of the brain and in psychopharmacology demand attention by social scientists, if only at the public policy level" (1992, p. 16). Although inclusion of the biological sciences in the social sciences is by no way an original theme (see Blank, Caldwell, Wiegele, & Zilinskas, 1998; Wiegele, 1979), dramatic advances in genetics and neuroscience over the past decade make it more imperative.

While it is vital that biology be incorporated into explanatory models of behavior, this does not mean that conventional factors be neglected. Although the biological bases of behavior, both genetic and neuronal, should not be viewed as deterministic forces, however, any model that diminishes their cumulative impact is simplistic and deceptive. Neuroscience, then, can be seen as a bridge between the natural and the social sciences (Cory, 1999). This does not require a reductionist program that shrinks all social science to fit the terms of biology as some early sociobiologists argued (see Losco, 2011), but rather one which serves to establish moorings and linkages through which the social sciences can introduce entirely new variables from their distinctive perspectives and levels of analysis. Even this is difficult because, as discussed in Chapter 1, biological-based models of behavior, even when heavily moderated by environmental influences, will always be controversial as they clash with the SSSM.

HEALTH STATUS AND NUTRITION

Although most recent research has centered on genetic and neural influences on behavior, human bodies and their functioning are an integral part of everyday life. While a broad array of physiological factors has been studied, systematic analysis continues to be lacking (Strate, 1991). Among those factors that have been investigated are health status, height and weight, age at puberty, menstrual cycle, body image, biorhythms, and psychophysiological arousal. Not surprisingly, there is substantial evidence that illness, poor health, and disability can have a profound impact on behavior. When illness strikes, a person is less able to carry out his or her usual roles and less able to engage in outside activities (Abeles & Ory, 1991). As a result, one adopts an illness role which defines appropriate behavior for a person in that condition (Sanders, 1982). Adopting this role protects the person from the criticism that he/she is not fulfilling normal expectations – as ill they are considered exempt from their normal obligations (Peterson, 1991). Moreover, adoption of the illness role is associated with increased passivity, exemption from carrying out normal social obligations, and a reduction in social functioning. Increased social isolation, lessened social interaction, depression, and lowered psychological well-being and sense of competence are also related to the illness role (Mechanic & Hansell, 1987).

Given that health status has direct and substantial effects on all aspects of a person's life, political behavior at some level is affected. One early study of rural elderly, for instance, found that poor health was correlated with lower levels of political efficacy, interest, and participation (Peterson, 1990). Other studies have found that poorer health is associated with depressed levels of participation and more negative and passive political views across a spectrum of populations, including students, adults, members of Congress, and federal appeals court judges (Schwartz, 1976). Closely tied to health status are physical fitness and energy levels. The more physically fit one is, the healthier that person is likely to be. Similarly, the healthier and more fit one is, the likelier he/she is to have higher energy levels. Unfortunately, the studies here have yielded inconclusive results, possibly due to methodological limitations. Most of these studies have relied on self-reports of fitness and energy level or on crude measures such as a treadmill test (Wiegele, Plowman, & Caty, 1975) or chest expansion (Booth & Welch, 1976). Unfortunately, the lack of significant funding appears to have discouraged more extensive and recent studies in these areas.

One of the few studies on the impact of health status on political behavior is that of Peterson (1990) based on a sample of rural elderly and two national samples. In addition to testing the health–behavior relationship, Peterson also hypothesized that healthier people would be more conservative in terms of party identification, acceptance of traditional values, lower support for government involvement in individual's lives, and self-identified conservative identification. Moreover, he suggested that healthier people ought to be less alienated. He concludes that "As health problems increase, political interest will decline, political efficacy will decline, political participation will decline, conservatism will decline, and alienation will increase" (1990, p. 86). While health status is not the dominant factor in shaping people's political orientations, it does have "an independent impact on a range of politically relevant variables" and should not be excluded as a predictor of political orientations and behavior (1990, p. 92).

As with health status, there is considerable research linking nutrition with behavior and there is reason to believe that it affects political behavior as well, despite the dearth of substantial political research. There are several reasons to suspect this linkage. Dietary deficiencies are associated with heightened depression, lowered energy level, apathy, irritability, impaired motor activity, and diminished cognitive ability (Liu, Raine, Venables, & Melnick, 2004). The principal impacts of malnutrition on political behavior are a function of lower energy levels, diminished mental and physical abilities, and apathy and depression. These biological effects lead in turn to decreased levels of political interest and activity (Schubert, 1981). Malnutrition in the first few years of life leads to antisocial and aggressive behavior throughout childhood and late adolescence. "Poor nutrition, characterized by zinc, iron, vitamin B and protein deficiencies, leads to low IQ, which leads to later antisocial behavior ... These are all nutrients linked to brain development." Researchers also found that the more the indicators of malnutrition there were, the greater the antisocial behavior (Liu et al., 2004).

Peterson argues that poor nutrition also affects people through its role as a stress producer because it diverts their attention from broader concerns as they try to cope with their personal problems. Constant hunger and inadequate nutrition under certain circumstances can act as a frustration and lead to heightened aggressive behavior, although comparative national research demonstrates this might be a transitional stage and is likely moderated by how repressive society is (Schubert, 1982).

In a study of older Americans, Petersen (1987) found that poor nutrition is correlated with diminished political efficacy and participation. In his more extensive analysis, he again found that poor nutrition, as expected, is associated with lower levels of political efficacy, political interest, and political participation. He also found that it is related to less conservative attitudes and more cynicism about politics. As with health status, the relationship between nutritional status and various political variables holds even when controlled for age, sex, income, and education. While the impact of nutrition on behavior is not huge, its nonrecognition in the political behavior literature is unwarranted.

Health and nutritional status have clear political ramifications (Strate, 1991). Populations whose health and nutritional needs are unmet are less politicized, less likely to approve of the status quo, and less efficacious. As such, poor health and nutrition help maintain the powerlessness of the disadvantaged and reduce their potential for political effectiveness. With this knowledge, it is possible that regimes can explicitly use biological oppression to pacify the masses. There is evidence that some authoritarian regimes manipulate food supplies as a biochemical means of political control. This might take the form of active starvation of rebellious populations into passivity or simply through neglect. "Hunger and poor health can be used to try and keep the powerless powerless, the disenchanted disenfranchised" (Peterson, 1990, p. 104).

HEALTH, DISABILITY, AND LEADERS

If biology and health influence mass behavior, it follows that the behavior of leaders can also be in part the product of biological factors. Unlike the scarcity of research on illness and mass behavior, public attention has at times focused on the impact of health on the behavior of particular leaders and there has been considerable debate as to whether persons with major health problems should hold public office. Questions of mental health and disability are raised during political campaigns and candidates make every effort to validate that they are physically and mentally fit for office.

Among the most likely physiological conditions affecting elite behavior are general health, physical or mental illness, stress, fatigue, alcohol, drugs, and hormonal imbalance. In contrast to the general lack of awareness of the public regarding the impact of health factors on its own behavior, there has frequently been widespread press coverage and public debate when applied to specific leaders, and some elections have been influenced by revelations of purported health problems. For instance, Senator Thomas Eagleton was dropped as vice-presidential candidate from the Democratic Party ticket in 1972 after revelations of past mental health problems. Also, Bill Bradley's quest for the Democratic Party nomination for President in 2000 effectively ended when it was disclosed that he had heart problems, even though they were minor and unlikely to physically interfere with presidential duties.

Likewise, historians have contemplated the extent to which health problems of figures such as Woodrow Wilson, Franklin Roosevelt, John Kennedy, Winston Churchill, and Adolf Hitler, among others, might have influenced history. According to Abrams (1999), 14 of the 18 U.S. presidents in the 20th century had significant illnesses while in office.

The episodic and individual-centered nature of the interest in health and leaders aside, few political scientists have shown interest in applying evidence from the life sciences to the study of leadership. As is the case in much of biopolitics, the work here has largely been concentrated in a small group of highly competent but relatively isolated researchers. Political scientist Robert Robins and medical doctor Henry Rothschild have written seminal articles on disabilities of presidents (1981) and about the responsibilities of president's physicians (1988). Other areas where scattered work has been published relates to how alcohol abuse affects decision making (Glad, 1988), how illness affects decision making (Gilbert, 2011, 2012; Park, 1986), and how dementia and aging might influence the ability to make decisions. The disclosure of Ronald Reagan's Alzheimer's disease after leaving office raised concerns over the extent to which this progressive disease affected his performance during his second term in office (Park, 1988).

Hugh L'Etang's work on the effect of drugs on political decision making (1988) provides a good base for more in-depth and systematic analysis and demonstrates through case studies of world leaders, the troubling situations it raises. Such research is even more essential in light of the ever-expanding range of psychotropic drugs available in the coming decades (Blank, 2013). Moreover, a biopolitical understanding of leadership itself, not just world leaders, should include an analysis of how biological variables – evolutionary, genetic, and neurological – influence the behavior of individuals in positions of authority. According to Wiegele, it is "at the individual level that knowledge from the life sciences will have its greatest impact on our understanding of how political elites, *as human beings*, function in decision-making situations" [emphasis added] (1979, p. 71).

Another area that has sparked some interest but which needs more methodical analysis is the impact of the public office on the officeholder. Most have heard the statement that the Presidency prematurely ages the occupant, and that presidents in general show the cumulative effects of a high stress job. But what scientific evidence is there? If so, is public office more or less stressful than corporate leadership? Are offices other than the national leader similarly stressful? Are persons who achieve high office more capable of adapting to a high stress environment on the basis of their biology? Are they somehow psychologically predisposed to seek out high stress positions?

At what stage does stress adversely affect decision making? What can and should be done to reduce the stress? None of these questions have been thoroughly studied, even though we live in an increasingly stressful world in many ways. They also lead us into the second dimension of elite behavior: are leaders biologically different than followers and if so how?

BIOLOGICAL BASES OF LEADERSHIP

Based on what was discussed earlier in this chapter regarding the linkages between our evolutionary roots, genes and brains and our behavior it would not be surprising if there was a robust biological influence on which persons in society seek to be leaders and ultimately achieve it. What specific traits it takes to be an effective leader of course depends heavily on the specific situation (it is generally understood that a person who might be a good leader under one set of circumstances might be a disaster under different conditions) as well as learned skills. The debate over the appropriate mixture of natural and nurtured leadership capabilities has been with us at least since Plato posed this question in *The Republic*, but unfolding understanding of neuroscience and genetics promises to support a natural dimension of leadership.

The literature and methods of ethology and related disciplines on dominance, authority, and attention structure have significant contributions to make to the study of elite human behavior, and early on there was considerable effort by a handful of first-generation researchers to apply these methods and perspectives. In her path-breaking work on authority among preschool children, Carol Barner-Barry (1977) used two techniques borrowed from ethology, nonparticipant observation and event sampling (Barner-Barry, 1996). These techniques allowed her to examine asymmetric authority interactions among the children and thus identify those children who exercise authority and those who did not. She found that even at preschool age, in the absence of adults, some children easily assume authoritative roles for the group for the purposes of interpreting and enforcing the rules. Interestingly, authoritative children intervened in conflict by exhibiting threat displays similar to those found in primate studies. Moreover, in most cases, only the authoritative children exhibited nurturing behavior.

In separate studies, Barner-Barry (1978) and Masters (1978) utilized the concept of attention structure, manifested primarily through visual awareness. Attention was operationalized by determining who is looking at whom and with what frequency. Overall, this research found that attention structures in stable groups of young children were quite similar to those exhibited in certain types of nonhuman primate groups. In his study of the emergence of leadership in young adults, Ivers found traditional explanations of elite socialization insufficient. He concluded that "more tightly focused ethological research in this area could have a profound impact on our understanding not only of leaders and their socialization, but also of the political socialization process more generally" (1997, p. 224).

Roger Masters has been instrumental in introducing a variety of experimental techniques, drawing on human ethology, social psychology, and cognitive neuroscience, to the study of leadership. His work on nonverbal displays, trait attributions, emotion, and cognition was instrumental in advancing the study of leadership and of political behavior in general (see Masters & Way, 1996). According to Sullivan and Masters, the extension of a biopolitical approach in this area has a dual effect. First, it adds the dimension of nonverbal cues and emotion to "the traditional focus on verbally expressed cognition that has long dominated political science," and second, "it permits a direct exploration of the way that species-specific nonverbal cues influence viewers' emotions" (1994, p. 238). In another study, Masters and Sullivan provide cross-national findings on nonverbal behavior and leadership (1989). Also see related work in Masters and Sullivan (1993), Sullivan and Masters (1994), and Salter (1996).

Other research related to leadership that make use of biopolitical approaches include Mazur and Mueller (1996), who examine how facial dominance provides an advantage in ascendancy to leadership positions in a hierarchy. In their study, Schubert, Wiegele, and Hines (1986) and Schubert, Wiegele, and Hines (1987) used a quasi-experimental field research design to study small group behavior in 24 city councils in three American states. They observed over 250 meetings involving over 500 hours of direct observation of conversational behavior in an effort to gauge the impact of age and age structure on political decision making. They found age to be an important though not determinative factor. Older council members were much stronger in their performance than younger members and older groups tended to be more cautious. Elsewhere, Schubert used ethology-based techniques to study verbal, visual, and vocal aspects of Bill Clinton's first inaugural address (1996). As with the previous researchers, Schubert goes beyond *what* is said and focuses on voice quality and nonverbal cues.

Early research into the biological bases of elite behavior also examined birth order (Forbes, 1971), hemispheric dominance (Kitchen & Peterson,

1996), handedness (LaPonce, 1976), and energy levels (Wiegele et al., 1975). Kitchen (1986), for instance, analyzed the language used by Richard Nixon and other Watergate figures to determine if they were reflecting left or right hemispheric processing. He found that the neurolinguistic patterns of Nixon, Haldeman, and Erlichman were strikingly similar in their reliance on language that draws on right hemispheric operations, while John Dean was more committed to left hemispheric speech patterns. In their study of birth order, Somit, Arwine, and Peterson (1996) reviewed a cross-national literature of nearly 2,000 publications on the birth order of leaders and concluded that there is no consistent significant relationship between birth order and leadership in Britain, the United States, or Soviet Union. Elsewhere (Somit, Arwine, & Peterson, 1997), however, they did find a definite first-born affect among female U.S. federal judges.

Judges were the subjects of one of the most extensive attempts to bring ethological and biosocial research methods to the study of elite behavior. Glendon Schubert's (1985) seminal work on the open deliberations of the Swiss Federal Tribunal paved the way for the oral argument project of the U.S. Supreme Court, which led to numerous papers and articles by the coinvestigators (see Schubert, 1997, for a description of these). Most observers of judicial behavior focus on written briefs at the exclusion of oral arguments. Although this makes for easier analysis, according to members of the Court itself (Harlan, 1955), this exclusion is a mistake. The oral argument study was designed to be the first systematic study of the substance and process of Supreme Court behavior utilizing acoustical as well as verbal record of the participant's speech. Schubert notes that nonverbal observation would have added significantly to our understanding of the dynamics of the process, but was not possible because of the Court prohibition on observers. With funding from the National Science Foundation, the project led to a wide range of publications (see Schubert, Peterson, Schubert, & Wasby, 1992). Sixty-four conclusions that emerge collectively from these studies are found in Schubert (1997).

Unfortunately, as with much elite-related behavioral research, despite intriguing findings, there has been little follow-up to most of these studies. Hopefully, more recent research on appearance effects in candidate appraisal (Schubert et al., 2000), sex and leadership (Ivers & Balash, 1999; Masters & Carlotti, 1994), and universals of command (Salter, 2000) will eventually be replicated in different contexts and extended to related areas in order to build a systematic database, although current attention on genetics discussed later in this chapter is likely to divert attention from these types of empirical research.

SOCIAL AND CULTURAL NEUROSCIENCE

In the past several decades, social scientists have made considerable inroads in understanding the mechanisms of human social interaction and cooperation (Kosfeld, 2007). Under the SSSM and the accompanying black box assumption, the focus was on the external social, environmental, and institutional determinants, while the internal biological mechanisms that originate and regulate individual decision making were generally ignored or dismissed as irrelevant. With the emergence of fields such as social neuroscience and neuroeconomics, however, neuroscientists have begun to look inside the brain and offer a more inclusive and detailed view of it. They have merged methods from the social sciences and cognitive neuroscience in an effort to understand what role the human brain plays in generating decisions. According to Theodoridis and Nelson (2012), although it is still a lesser movement than its equivalents in economics and psychology, social neuroscience is generating a wealth of information about the location of neural activation when subjects perform various tasks or are presented with certain stimuli and it has far-reaching implications for the study of political behavior.

Social neuroscience investigates the biological mechanisms that underlie social structures, processes, and behavior and the interactions among social and neural structures and processes (Lupia, 2011a, 2011b). It is challenging because it necessitates the integration of multiple levels of analysis from the genome to social groups and cultures. Mapping across systems and levels requires interdisciplinary expertise, comparative studies, innovative methods, and integrative conceptual analysis (Cacioppo & Decety, 2011).

Since human beings are fundamentally social creatures whose mental and physical health depends on their social position and relationships, the human brain has evolved to promote social coordination, communication, interaction, and collective enterprises (Cacioppo & Decety, 2011). Living in a social group is a major adaptive advantage, but it presents new challenges. To get along and get ahead, it is necessary to know who is friend and who is foe. Also, it can be advantageous to form alliances with some group members in one context but to outmaneuver them in another. The social brain hypothesis suggests that, from an evolutionary standpoint, living in bigger, complex social groups selected for larger brain regions with a better capacity to perform relevant computations (Weaver, 2010).

A complementary field, cultural neuroscience, attempts to understand how culture as an amalgam of values, meanings, conventions, and artifacts interacts with the mind and its underlying brain pathways

(Kitayama & Park, 2010). According to Chiao and Ambady (2007), the use of neuroimaging and event-related potential techniques to study the effects of culture on the spatial and temporal features of the neural response to perceptual, cognitive, social, and emotional tasks is integral to cultural neuroscience. This research has also led to neuroanthropology, a humanistic science for the study of the culture–brain connection, which Dominguez Duque, Turner, Lewis, and Egan (2010) contend can make a significant and distinctive contribution to the study of the relationship between culture and the brain.

Until recently, the relevance of culture for brain development, structure, and function, and vice versa, was unrecognized, but this has changed with the integration of cross-cultural psychology, cognitive neuroscience, and molecular biology to examine how brain structure and function vary from one cultural group to the next (Han & Northoff, 2008). The first generation of cultural neuroscience studies has yielded a number of important insights. First, culture seems to have a pervasive effect on all levels and dimensions of neural activity, from low-level perceptual and attention processes to high-level cognitive, affective, and social functions, including activity associated with language, music, mental calculation, emotions, mentation, and self-awareness (Seligman & Brown, 2010). In sum, differences in cultural systems are reflected in disparities in brain systems.

Second, cultural differences appear to be reflected in different arrangements of neural circuits or in different levels of activation. Thus, cultural experiences not only modulate but also govern preexisting patterns of neural activity and are thus constitutive of that experience. There is preliminary evidence that cultural differences can be expressed in structural changes in the brain (Han & Northoff, 2008). For instance, Gutchess et al. (2010) report that during categorization, East Asians recruit regions in the frontoparietal network, whereas Westerners recruit the temporal lobe and the cingulate cortex to a greater extent. Moreover, in a comparison of Caucasian American and Korean subjects, Cheon et al. (2011) found that cultural variation in preference for social hierarchy led to cultural variation in in-group preferences in empathy resulting from an increased engagement of brain regions associated with representing and inferring the mental states of others.

While fruitful work has focused on how culture influences the contents of cognition, culture can also exercise a profound effect on the *how* of cognition – the mechanisms by which cognitive tasks get done. Many fundamental processes of daily cognitive activity involve the operation of cognitive tools that are not genetically determined but rather invented and culturally transmitted. Further, these cognitive creations become firmware, thereby constituting a reengineering of the individual's cognitive architecture by his or her cultural context, resulting in a variety of cognitive phenotypes (Wilson, 2010). There is evidence that humans possess a unique, cooperatively based social cognition that allows an especially rich form of social learning and cultural transmission (Warneken & Tomasello, 2009). While cultural accumulation cannot explain human cognitive universals, it is, nevertheless, a highly important source for determining the adult cognitive phenotype. While brain regions include strong predispositions to perform certain cognitive functions, there is a considerable "fringe of variability" (Dehaene, 2005) that allows each individual's experiences and learned cognitive habits to rewire the system.

Losin, Dapretto, and Iacoboni (2010) conclude that the combination of culture and neuroscience is additive and synergistic and that cultural neuroscience studies have the potential to transform existing methodologies and produce unique findings. They argue that one of the most productive areas of cross-cultural comparison is social cognition. A commonly used classification of cultures is the dichotomy individualistic or collectivistic. Individualism is often characterized by the assumption that individuals are separate in identity and responsibility from one another and is typically associated with Western cultures, particularly the U.S. Collectivism is characterized by the idea that individuals are defined by and obligated to their social groups and is typically associated with Asian cultures. Losin et al. (2010) conclude that future investigations in cultural neuroscience will allow us to investigate the neural correlates of individualistic and collectivistic orientations.

Human experience is patterned by culture, and the human mind is likely shaped by cultural variables. These culturally patterned behaviors and experiences can, in turn, influence the connectivity of the brain itself. Indeed, the brain is the principal site for cultural information to accumulate throughout one's life. Such cultural values contain meaning and encompass several elements, including values, practices, and psychological tendencies. This process does not exist in isolation, but rather produces a reciprocal loop which itself contributes to the production and change of cultural practices and identity formation and enculturation. This interaction results in neuroplastic changes in the brain throughout one's lifespan in response to experience, offering affirmation of self-identity and well-being as well as resulting in other biological changes (McDermott, 2011a, 2011b, p. 99).

A voluminous amount of research on identifying the neural structures involved in processes of social cognition and about the neural processes defining the role of affect and emotion in cognition, perception, and memory has emerged since 2000. Although much of this research is peripheral to the study of political behavior per se, this "fast-developing branch of social neuroscience has already spread into political science in certain ways" (Theodoridis & Nelson, 2012, p. 31). This section provides an overview of some of the most relevant areas for the study of group behavior.

Herding in Humans

One concept for comprehending group behavior is herding, which is defined as the alignment of the thoughts or behavior of individuals in a group through local interaction without centralized coordination. Herd mentality and behavior have been a ubiquitous part of humans since individuals began to construct tribes and migrate collectively (Raafat, Chater, & Firth, 2009). Although most interest in herding has focused on stock markets, herding theory has relevance for many areas of politics, from voting behavior to political protests and mob violence, and is reflected in such concepts as mass hysteria, the bandwagon effect, and group-think. For example, several studies have shown that happiness and obesity tend to spread through social networks in a manner analogous to a contagious disease (Fowler, Dawes, & Christakis, 2009). Novel neuroimaging techniques that scan many individuals at a time could be used to capture brain-based correlates of herding, while neuroeconomics offers the possibility of characterizing and building biological models of herding (Sanfey, 2009). One potential research area on susceptibility to herding is the impact of social networking and its increased capacity for sharing information immediately to ever-proliferating points of contact for an individual (Raafat et al., 2009).

Social Trust

At the core of social relationships is the concept of social trust. Trust pervades human societies, is indispensable in friendship, families, and organizations, and plays a key role in economic exchange and politics (Kosfeld, Heinrichs, Zak, Fischbacher, & Fehr, 2005). In the absence of trust in a country's institutions and leaders, political legitimacy breaks down. While trust can be a risky decision because the trusting person risks being exploited, the hope is that trust will be rewarded by reciprocal behavior, thereby increasing the efficacy of the social interaction (Kosfeld, 2007). Thus, the economist Kenneth Arrow described trust as "an important lubricant of a social system" (Arrow, 1974). However, since the risk involved in the trusting decision can never be fully resolved, the trusting person must always overcome his or her aversion to the risk of being exploited in order to achieve maximal benefits from the social interaction.

Although research has identified many conditions affecting the level of trust, such as recurrent interaction, the possibility of sanctions, and the activation of social norms, the question remains as to what is it in the brain that makes humans trust each other. Koscik and Tranel (2011) hypothesize that damage to the amygdala can interfere with normal patterns of interpersonal trust and result in increases in reciprocity despite betrayals of trust. In addition, researchers have identified oxytocin as promoting increased levels of trust. Oxytocin, a neuropeptide produced in the hypothalamus and stored in the posterior pituitary, is naturally released into the bloodstream and widely distributed throughout the CNS, but it can also pass through the blood–brain barrier through intranasal administration. Oxytocin reduces anxiety and acts as a central regulator of stress-protective social behavior. It has also been shown to enhance the willingness to trust others and accept social risks arising from interpersonal interactions (Baumgartner, Heinrichs, Vonlanthen, Fischbacher, & Fehr, 2008). In a study by Kosfeld (2007), subjects received either a single intranasal dose of oxytocin or a placebo. Oxytocin increased subjects' willingness to trust their trustee, although it did not make them more optimistic about the latter's trustworthiness (De Dreu, Greer, Handgraaf, Shalvi, & Van Kleef, 2010).

The discovery that oxytocin increases trust in humans has important implications for patients with mental disorders associated with social dysfunctions such as social phobia or autism. At the same time, these findings have provoked fears of potential abuse, ranging from unscrupulous employers or insurance companies using it to induce trusting behavior in their employees or clients to fraudulent salesmen spraying customers. Some suggest that governments could attempt to sway public opinion or stifle dissent by chemically enhancing citizens' trust in its actions or that political candidates could use it to gain unearned trust at the polls (Penney & McGee, 2005). According to Kosfeld (2007), however, these fears are groundless because the surreptitious administration of an effective dose of oxytocin is technically impossible and forced nasal administration could raise the recipient's level of distrust, therefore negating the intended effect.

Empathy

Highly related to trust for understanding both individual and group behavior is the concept of empathy, which involves the capacity to share the affective experiences of others. This ability to share others' feelings results in a better appreciation of the mental states and actions of others and thereby promote prosocial behavior (Singer & Lamm, 2009). Philosophers have long conjectured about empathy, but recent research has begun to provide a better grasp of why, when, and how we experience empathy and whether we can use that knowledge to increase prosocial behavior and reduce conflict (Decety, 2010). As with trust and altruism, there is strong evidence that empathy has deep evolutionary, neurological, and biochemical underpinnings. More advanced forms of empathy in humans are built on basic forms and remain connected to core mechanisms associated with affective communication, social attachment, parental care, and motivation to help others (Cacioppo & Decety, 2011).

Most studies of empathy utilize the observation of pain in others to evoke empathic responses (Lamm, Meltzoff, & Decety, 2010). Many fMRI studies have demonstrated striking similarities in the neural circuits involved in the processing of both the first-hand experience of pain and the second-hand experience of observing it in others. One common finding is that perceiving pain in others activates the same neural networks that are triggered when we experience pain ourselves. According to Singer et al. (2004), empathy works by tapping into a brain mechanism that already exists for our own pain, thus making us believe we are feeling pain emotionally even when we are not feeling it physically.

Many studies have found that females have on average a stronger drive to empathize, while males on average have a stronger drive to systemize (Baron-Cohen, 2010). Singer et al. (2006) concluded that when it comes to deriving pleasure from someone else's misfortune, men seem to enjoy it more than women. Using fMRI, they compared how men and women reacted when watching other people suffer. If the sufferer was someone the subject liked, areas of the brain linked to empathy and pain were activated in both sexes, but only women had a similar response if they disliked the sufferer, while men instead exhibited a surge in the reward circuits. Thus, women had a diminished but still active empathic response if they disliked the target, whereas for men it was completely absent. Importantly, empathic responses to other people, particularly in men, are not automatic but depend on the emotional link to the suffering person. Men also admitted to having a much higher desire for revenge and derived satisfaction from seeing the unfair person being punished. Similarly, Ma et al. (2011) found more empathy toward friends than strangers and concluded that both familiarity and self-engagement are factors that influence empathy. Furthermore, Cheon et al. (2011) suggest that cultural variation in preference for social hierarchy leads to cultural variation for in-group preferences in empathy because of the increased engagement of brain regions associated with representing and inferring the mental states of others.

Shirtcliff et al. (2009), however, propose that neurobiological impairments in individuals who display little empathy are not necessarily the result of a reduced ability to understand the emotions of others but instead are connected to decreased physiological arousal to their own distress. One manifestation of this reduced stress reactivity may be a dysfunction in empathy leading to callousness. Callousness and unemotional (CU) traits are related to maladaptive social information processing (Frick & White, 2008). While empathy can promote affiliation and prosocial behavior, CU traits are associated with antisocial behavior and are particularly useful in predicting which antisocial youth will continue offending into adulthood (Shirtcliff et al., 2009). Deficits in emotionality, disruptions in empathy, and the failure to respond to the distress cues of others (i.e., callousness) are at the core of the impaired decision-making capabilities (Kimonis et al., 2008).

In his book, *Zero Degrees of Empathy*, Baron-Cohen (2011) sees "evil" as the product of a failure to empathize. He terms psychopathic personality disorder, borderline personality disorder, and narcissistic personality disorder "zero negative" because affected subjects' lack of empathy is considered to be solely detrimental, and autism spectrum disorders "zero positive" because, while affected persons lack empathy, they also have systematizing natures that are valued because of their focus on detail and patterns. He attributes these traits to dysfunction in the brain's empathy circuit, genetic abnormalities, and poor developmental conditions that preclude the forging of an emotionally resilient psychology normally developed through secure attachment. Baron-Cohen contends this helps explain human cruelty, since humiliation can be used to establish status, to signal collaboration with a dominant person, or to respond to one's own perceived oppression. Additionally, dehumanization may arise when perpetrators have to evade their intact empathy mechanisms in order to permit murder via indirect methods, such as by mercenaries, through the use of impersonal technologies such as drones, or by forcing victims to kill one another. According to Preston (2011), only through an interdisciplinary framework that combines neuroscientific knowledge with findings from social and political science can we grasp the richness of the human context under such inhumane circumstances.

Aggression and Fear Center

One challenge endemic to all societies is to better comprehend and curtail the sources and forces of aggressive and violent behavior. "As a result, it seems that one of the most important areas in which the natural and social sciences might converge for effective collaboration surrounds the neurobiological and neuropsychological bases of aggression and violence" (McDermott, 2011a, p. 100). Although emerging knowledge in genetics and neuroscience clashes with current assumptions about behavior and cognition in general, perhaps the most significant political findings are those bearing on aggression and a constellation of affective antecedents and behavioral parallels, including fear, anger, hatred, anti-sociality, impulsivity, criminal behavior, and violence (Eisenberger, Way, Taylor, Welch, & Lieberman, 2007). No credible scholar today would argue that the causes of aggressive behavior are either wholly environmental or wholly biological. Although in rare cases, aggression might be essentially biologically or environmentally based, in a preponderance of cases it cannot be attributed to any single factor (Garrett & Chang, 2008). The most appropriate approaches, therefore, are those that explicate how a complex set of biological factors interacts with and influences a mixture of environmental factors to produce an aggressive behavior (Díaz, 2010). Mehta and Beer (2010), for instance, view aggression as a complex social behavior that is regulated by multiple social and biological factors working together as parts of an integrated system.

Furthermore, the line between biological and social or environmental factors is obscured because many brain deficits linked to aggression are, themselves, the product of environmental insults. For example, it has long been known that alcohol elicits aggressive and violent behavior in some people. Similarly, environmental insults are capable of producing life-long injuries to the developing brain. For instance, lead exposure in children is extremely risky for neural development and supported by hair analyses of serial murderers and violent offenders that reveal excessive concentrations of lead and cadmium. Masters and Coplan (2001) also found that when lead and manganese interact, the deleterious effects are magnified because they compromise the serotonin, dopamine, and other neurotransmitter systems that are integral to self-control and setting thresholds for violent behavior.

To date, most attention has focused on hormones, neurotransmitters, the limbic system, and the "fear circuit" in the prefrontal cortex (PFC) (Rodrigues, LeDoux, & Sapolsky, 2009; Wahlund & Kristiansson, 2009). A credible theory is that violent impulses originate deep in the limbic system, amygdala, and the hypothalamus, and that the job of the higher brain regions, particularly the PFC, is to moderate these impulses (Wahlund & Kristiansson, 2009). Therefore, lesions in the PFC or disrupted functioning of the neurotransmitter serotonin that is central for impulse control undercuts its capacity to serve as an inhibitory brake on the fear center and properly regulate social behavior (New et al., 2007). Although specific genetic or neuronal patterns that explain aggression have not yet been found, certain anomalies that predispose a person to impulsivity and uncontrollable aggression have been found (Glenn & Raine, 2009; National Institute on Drug Abuse, 2010). For instance, traumatic brain injury can lead to aggressive behavior (Reeves & Panguluri, 2011). PFC damage caused by birth injury, childhood infection, neurotoxin exposure, or brain trauma is also correlated with hyperactivity and impulsiveness that can lead to aggressive behavior (Hoptman & Antonius, 2011). Moreover, early developmental damage in this region is associated with impairment of social and moral reasoning, as is common in psychopaths (Schmook, Brugger, Weber, & Kasprian, 2010).

Zaretsky, Mendelsohn, Mintz, and Hendler (2010) suggest that this brain network is involved in generating subjective assessment of social affective cues. The amygdala is activated most strongly when lacking sufficient data regarding a potential threat in the surroundings (Hsu, Bhatt, Adolphs, Tranel, & Camerer, 2005). Especially in ambiguous situations, the amygdala may act to decrease the threshold of relevant sensory systems in order to gather additional information for resolving the ambiguity of the signal and increasing its predictability. If this information resolves the uncertainty, amygdala activity subsides. In this way, the amygdala plays a crucial role in formulating a predictive decision under uncertainty. "If its threshold is set too low, normally benign aspects of the environment are perceived as dangers, interactions are limited, and anxiety may arise. If set too high, risk taking increases and inappropriate sociality may occur" (Schumann, Bauman, & Amaral, 2011, p. 745).

Imaging studies have reported greater activation of the amygdala in response to emotional facial expressions, especially fear, indicating that it is capable of rapidly processing fearful and sad facial expressions (Fenker et al., 2010; Kim et al., 2010). Pictures of fear-provoking faces initiate a quick rise and fall of activity in the amygdala (Sato et al., 2011). Moreover, Olsson, Nearing, and Phelps (2007) found that fear acquired indirectly through social observation with no personal experience of the aversive event engages similar neural mechanisms as fear conditioning. In contrast to empathy, which facilitates sharing the affective experiences of others, fear, and related emotions are at the base of in-group versus out-group tensions. Importantly, the fear process is largely beyond conscious control (Custers & Aarts, 2010).

This emphasis on the role of the amygdala does not negate the complementary roles of a broad network of regions, including the medial prefrontal and anterior cingulate cortices, in mediating limbic-initiated fear and generating emotional responses. Davidson, Putnam, and Larson (2000) posit that while normal individuals have the capacity to regulate their negative emotions, aided by restraint-producing cues such as facial signs of anger, persons predisposed to act aggressively have an abnormality in the central circuitry responsible for these adaptive strategies. They conclude that the propensity for reactive (impulsive) aggression is associated with a low threshold for activating negative affect and a failure to respond appropriately to the anticipated negative consequences of one's behavior. Practically speaking, such findings have clear implications for how we react to others in a social setting. The neurological foundations of hate or ideology remain unsettled, but "as surely as there are neurobiological mediators of aggression, there are neurobiological mediators of ideology" (Perry, 2002).

In addition to the contribution of deficits in regions of the PFC and the anterior cingulate cortex, chemically there is evidence that anterior brain dysfunction in aggressive individuals involves disruptions in serotonergic function (Seo, Patrick, & Kennealy, 2008). Abnormalities of serotonin metabolism also have been found disproportionately in impulsively aggressive men, compulsive fire-setters, and the violently suicidal (Venables, Patrick, Hall, & Bernat, 2011). Moreover, aggression often begins when self-control stops. In turn, self-control requires considerable brain food in the form of glucose. Thus, people who have difficulty metabolizing glucose are at greater risk for aggressive and violent behavior (DeWall, Deckman, Gailliot, & Bushman, 2011).

Despite ongoing dispute over the link between testosterone and aggression, high levels of testosterone have been reported in populations of aggressive individuals (Siever, 2008). Mehta and Beer (2010) found that testosterone amplified aggressive behavior by reducing the ability of the medial orbitofrontal cortex (OFC) to govern self-regulation and impulse control following provocation, while Derntl et al. (2009) conclude that increased levels of testosterone improved the amygdala's ability to process threat-related stimuli. Furthermore, McDermott, Johnson, Cowden, and Rosen (2007) found that high-testosterone subjects were much more likely to engage in unprovoked attacks against their opponents than were their low-testosterone counterparts. Although female participants were as likely as the men to fight back once they were provoked, men were much more likely to initiate a conflict. Indeed, males between the ages of 12 and 25 are the principal perpetrators and victims of violence, which could be due to heightened testosterone levels that escalate in early adolescence (Craig & Halton, 2009).

With its central role in human adaptability and survival, it would be reasonable to expect that aggressive traits resulting from variation in brain function would be influenced by genes. According to O'Donovan and Owen (2009), there is convincing evidence that discrete genes are implicated in aggression. One hypothesis is that gene abnormalities result in structural brain abnormalities, which result in emotional, cognitive, or behavioral abnormalities, which in turn predispose one to antisocial behavior (Raine & Yang, 2006). Other studies suggest that stress may intensify this response (Price & Drevets, 2010). To date, at least seven genes meet the dual criteria of being associated with antisocial/aggressive behavior in humans or animals and influencing brain structure: MAOA, 5HTT, BDNF, NOTCH4, NCAM, tlx, and Pet-1-ETS (Raine, 2008). By far the most investigated is MAOA, a gene that codes for an enzyme which degrades dopamine, norepinephrine, and serotonin (Eisenberger et al., 2007). This gene has earned the nickname "warrior gene" because it has consistently been linked to aggression in observational and survey-based studies (McDermott, Tingley, Cowden, Frazzetto, & Johnson, 2009). Males with a common variant in the MAOA gene have an 8 percent reduction in the volume of the brain structures that are involved in emotion (Meyer-Lindenberg, Buckholtz, Kolachana, Hariri, & Pezawas, 2006) and those who are deficient in MAOA, consequent to an X-chromosome defect at the vicinity of the MAOA gene, demonstrate frequent aggressive behavior (Soliman et al., 2010).

In addressing the question of environmental influence, McDermott and colleagues (2009) concluded that aggression occurs with greater intensity and frequency as provocation is manipulated upward, especially among low-activity MAOA subjects. There is mixed evidence of a main effect for genotype and for a gene by environment interaction, such that MAOA is less associated with the occurrence of aggression in a low-provocation situation but significantly predicts such behavior in a high-provocation one. Furthermore, there is evidence that a functional polymorphism in the MAOA gene can mediate the impact of traumatic early life events on the propensity to engage in violence as an adult. Specifically, children who had suffered abuse and who had the low-activity form of MAOA were much more likely to develop antisocial problems as adults (McDermott et al., 2009).

There is also evidence that the dopaminergic system plays an important role in aggressive and antisocial behavior. Low levels of dopamine are associated with attention deficit/hyperactivity disorder, and the dopaminergic system also contributes to both aggressive and nonaggressive anger and delinquency. In turn, delinquency, anger, and impulsivity have strong genetic components (Ebstein, 2006). The dopamine receptor D4 gene (DRD4) located on chromosome 11 has received special attention for its potential impact on reward motivation, attention, and approach behavior (Couppis, Kennedy, & Stanwood, 2008). The findings for the DRD4 gene, however, have been mixed, implying that behavioral expression of the gene could be moderated by other factors. Indeed, several studies have found that exposure to a negative social environment exacerbates the risks associated with the DRD4 gene (Dmitrieva, Chen, Greenberger, Ogunseitan, & Ding, 2011). Moreover, as with MAOA, the link between DRD4 and anger has been confirmed for males but not females. This gender-specific vulnerability to DRD4 could be attributed to gender differences in biological factors, such as estrogen and progesterone, sex chromosome genes, or autosomal genes (Harrison & Tunbridge, 2008), or to differences in the exposure to contextual risk factors. Dmitrieva et al. (2011) conclude that the association between the DRD4 VNTR polymorphism and adolescent delinquency, short temper, and thrill seeking is stronger for males than for females, and that this gender-specific expression of the DRD4 gene can be explained by gender differences in exposure to violence and parental monitoring of youth activities.

Additionally, a number of studies have linked fear, ethnocentrism, and out-group attitudes with genetic influences (Hatemi & McDermott, 2012a). For instance, Hatemi, McDermott, Eaves, Kendler, and Neale (2013) found that individual distinctions in fear dispositions help explain differences in attitudes toward out-groups. Such fear dispositions are part of a person's genetically influenced psychological make-up and conditioned through specific developmental pathways. They intimately affect patterns of attachment and operate through genetic predilections that connect social fear to out-group attitudes. "Our findings identify an important mechanism by which genetic differences influence political attitudes and indicate distinctions that *social* fear might serve as a foundation for some part of the edifice of certain aspects of political ideology" (Hatemi et al., 2013, p. 179). Similarly, Stam, Von Hagen-Jamar, and Worthington (2012) found that a majority of the correlation between social fear and immigration attitudes was due to a common genetic factor. This suggests that while genes do not directly affect specific attitudes, genetic propensities influence the disposition and operation of an emotive condition, which in turn is manifested toward external targets, including strangers and out-groups, when elicited. This does not mean that social environments are unimportant or that genetic influence is fixed. Obviously, individuals learn who the relevant out-groups are through social learning and reinforcement. Nonetheless, sensitivity to fear of the out-group is in part genetically informed, and this genetic influence is manifested as anti-out-group attitudes on issues such as immigration and national security.

We are unlikely ever to fully understand, much less to predict or treat, aggression and violence by looking only at genes and the brain because at their base such behaviors are a product of political and social, not solely biological, problems (Mercadillo & Arias, 2010). As noted elsewhere, the critical question to ask is "how, not whether, biology and environment are related – how a complex of biological factors interacts with a complex of environmental factors producing a single isolated aggressive behavior or a patterned concatenation of aggressive behaviors" (Blank, 2006, p. 15). While a better knowledge of the role of the brain and human behavior is not sufficient to explain human aggression, "one cannot understand violence without a thorough understanding of the human mind" (Pinker, 2002, p. 317). In other words, although it is unlikely that we will ever be able to describe a particular aggressive act or even a type of aggression in terms of specific neuronal activity, it is crucial that this neural dimension be an integral part of any credible paradigm of behavior.

Implications for Policymakers

Clearly, there is considerable research on the neural foundations of bias and its ties to inter-group conflict. Imaging studies have found that even among seemingly unprejudiced people, racial category labels prime stereotypes and out-group cues such as faces activate negative evaluative terms. Moreover, research on the limbic system suggests that dependence on rational incentives is unlikely to resolve the fears and hatred among conflicting groups and that even the leaders of the opposing sides are limited in their capacity to convince their followers to accept peace without ameliorating emotional biases deep within the brain.

One question is how, as policymakers deal with an unpredictable and threatening world, they can best assimilate the emerging knowledge about the brain's contribution to aggression, bias, and inter-group conflict into foreign policy? Are there more effective ways than force or traditional mediation with handpicked leaders to resolve such problems? What good does it do to remove a ruthless dictator if his followers hold similar views of outsiders? Moreover, what dangers are there when leaders characterize certain easily identifiable ethnic groups as the enemy? Or when they profile based on physical characteristics? What deep-seated evolutionary fear circuits do such policies trigger? In contrast, to what extent do leaders exploit the fear system to build support for their policies, and what are the ramifications of such fear system engagement in the long run?

Emotion, Cognition, and Decision Making

At the core of political behavior and the center of politics is the ability to make and execute decisions. The study of decision making spans varied fields such as neuroscience, psychology, economics, political science, and computer science, but despite this diversity of fields and applications, most decisions share the common elements of deliberation, motivation, and commitment (Gold & Shadlen, 2007). Not surprisingly, an emergent area of research is decision neuroscience, which attempts to integrate research in neuroscience and behavioral decision making. Neuroscience has already made an important contribution by demonstrating that decision making involves the orchestration of multiple neural structures and cognitive systems. Research has shown that areas such as the ventromedial PFC, amygdala, insula, somatosensory cortex, dorsolateral PFC, and hippocampus are all involved in various aspects of decision making (Gupta, Koscik, Bechara, & Tranel, 2011).

According to Pessoa and Engelmann (2010), executive processes, such as attention, working memory, and inhibition, constitute a set of processes that are particularly important for behavioral planning and production. Since the brain, unlike a computer, has a restricted processing capacity, in order to efficiently guide behavior it needs to be able to segregate those stimuli that deserve further processing from those that are better left ignored.

To this end, both bottom-up and top-down processes are involved. By using top-down control, the brain can more efficiently allocate its resources based on prior knowledge and current behavioral goals. Concurrently, processing resources can preferentially shift to salient features of the environment. Based on behavioral evidence, both of these processes are intimately linked to reward and motivation. Furthermore, these findings complement previous demonstrations that the motivational dimensions of (top-down) goals rely on the dopamine system and its projection sites, while (bottom-up) stimulus salience is encoded in specific nodes of the reward system, such as the caudate nucleus (e.g., Zink, Pagnoni, Chappelow, Martin-Skurski, & Berns, 2006).

While the importance of emotions is frequently rejected or ignored in favor of theories that stress reasoning, decision making rests on an integrated notion of emotional rationality. As cogently stated by McDermott, "emotion is intertwined with cognition in a way that requires the processes to be analyzed interdependently; emotion is, inescapably, an essential component of rationality" (2004, p. 700). This, of course, is contrary to rational choice models that limit the role of emotions by maximizing personal benefits based on reason alone. As discussed earlier, rational choice assumes that humans are capable of making purely rational decisions. Therefore, it maintains that emotions should not play a major role in politics, and, in fact, that "uncontrollable passions" are dangerous (Hall, 2009). This highly pessimistic view of the emotions and the role they play in political motivation and decision making ignores philosophical debates over the proper role of the passions in politics throughout the history of political thought. Plato and, especially Aristotle, saw emotions as actively beneficial for political life, with reason and emotions engaging in a kind of symphonic relationship and emotions providing the attachment and motivating force that reason lacks (Sokolon, 2006). Similarly, Hume, Mill, Rousseau, and others have stressed the importance of emotions for political decision making.

Evolutionary theory and mounting evidence from neuroscience demonstrating that cognitive and emotional pathways are inextricably connected, however, are shifting political analysis toward a view that emotions can play a crucial role as positive components of political discourse at the center of decision making. As humans, our emotional reactions to developments and emotive connections to other individuals and society help define us. Emotions are the foundation of the social and political relationships that make human life at times trying, but ultimately meaningful. Therefore, according to Hall (2009), political scientists lose predictive power and scholarly relevance when they ignore the major normative debates over the role of the emotions in political life and the accumulating empirical research indicating the importance of impassioned discourse. Rather than dismissing emotion, understanding how it influences politics can help mitigate the negative effects of emotional appeals while explaining their role in motivating information seeking and the social attachments that countenance political discourse.

Emotional processing does not merely exert a negative impact on decision making, but rather constitutes an inherent part of rationality itself and facilitates quick, effective, and accurate decision making. Under normal circumstances, emotional responses offer important and useful information about other people, oneself, and external events (McDermott, 2004). Since it misses this major piece of the puzzle, it is not surprising that rational choice models have proven to have limited explanatory power or offer consistently good predictions about what citizens will do. Moreover, simple cost–benefit analysis seems to have limited explanatory value, in large measure because politics fundamentally engages values, beliefs, and norms, all of which are affected by emotion (Hall, 2009, p. 85).

An emotion is defined as a mental state that arises spontaneously rather than through conscious effort and is often accompanied by physiological changes. We often refer to emotion as a feeling, for instance, of joy, sorrow, reverence, hate, love, and so forth. The term emotion, then, commonly denotes a specific phenomenal state. The questions as to where such feelings originate and what role they play in human behavior have been at the core of philosophical debate. As affective neuroscience has begun to identify the neural bases of these mental experiences, one of the brain structures consistently associated with emotional function is the amygdaloid complex. It is believed that it is involved in coupling a stimulus which evokes an emotional response with its affective value (Gupta et al., 2011). As noted earlier, it is well established that the amygdala plays a key role in the processing of emotional information, regulates emotional responses, and controls fear reactions across a range of species (Adolphs, 2010; Modinos, Ormel, & Aleman, 2010).

The amygdala is interconnected with both cortical and subcortical structures and, thus, is part of many intersecting networks of regions that are activated while behaviors unfold (Pessoa, 2010a, 2010b). It is positioned to make a crude but rapid appraisal of the state of the world and alter the perceptual process by enhancing the encoding of events of potential importance for the individual. Simply put, it determines if information received by the senses represents a threat to survival (Pessoa & Adolphs, 2010). If not, it recognizes that there is time for more advanced processing through the PFC, which is responsible for more abstract, rational thought. However, if the information is considered a threat, the amygdala shuts down the higher mental processes through the release of catecholamine neurotransmitters, which trigger an almost immediate reaction by the emotional brain. This processing hierarchy presumably evolved to help humans and other species to survive fight-or-flight situations. Importantly then, emotional processing occurs before, and sometimes instead of, higher-level cognitive functioning. "This finding alone begins to undermine theories of rationality that presume that emotion is either not involved in decision making or exists in opposition to the highest-quality decision making" (McDermott, 2004, p. 693).

Another part of the brain that is critical for emotions is the OFC which is heavily involved in emotion, mood, drives, and rewards and, like other PFC structures, it mainly serves to regulate emotion, control mood, monitor rewards and punishments, and be engaged during planning and decision-making tasks. Together, the OFC and the amygdala promote stimulus-reinforcement learning and decision making in healthy individuals, particularly during associative learning (Schoenbaum, Saddoris, & Stalnaker, 2007). According to Adolphs (2010), this prefrontal activity is involved in downregulating activity in emotion-generation regions by its inverse association with the amygdala response to negative stimuli. Moreover, in their meta-analysis of emotion, Vytal and Hamann (2010) found that each of the basic emotion states (anger, fear, sadness, anger, and disgust) was associated with consistent and discriminable patterns of neural activation, and that the signature patterns of neural activation that characterized each emotion consistently differentiated that emotion from other emotions.

Emotion and Cognition

Historically, emotion and cognition have been viewed as largely separate processes executed by different regions of the brain, such as the amygdala for emotion and the PFC for cognition. One way in which emotion has been contrasted with cognition has been to link the former with "irrational" or "suboptimal" processes that are more basic, namely, more linked to survival, than cognitive ones (Pessoa, 2010a). Under this framework, functional interactions between the amygdala and the PFC mediate emotional influences on cognitive processes such as decision making as well as the cognitive regulation of emotion. More recent data suggest, however, that such view is misleading and that, in order to understand how complex behaviors are carried out in the brain, it is necessary to understand the interactions between emotion and cognition (Pessoa, 2010a).

In their meta-analysis of research on the representation and utilization of cognitive and emotional parameters, Salzman and Fusi (2010) contend that these mental state parameters are "inextricably linked" and represented in dynamic neural networks composed of interconnected prefrontal and limbic brain structures. Thus, cognition and emotion are considerably more effectively integrated in the brain than previously assumed. Emotional processes can influence cognitive processes and cognitive processes can regulate or modify our emotions. Both of these interactions can be implemented by changing mental state variables (either emotional or cognitive ones), and emotions and thoughts shift together, corresponding to the new mental state. For Pessoa (2009), the impact of an emotional stimulus on behavior depends on how it affects the flow of executive functions. He hypothesizes that the results will depend upon the level of threat, which will determine if emotional content enhances or impairs behavioral performance.

In their study, Camerer, Loewenstein, and Prelec (2005) conceptualize the two dimensions as shown in Table 1. The first dimension distinguishes between controlled and automatic processes. Controlled processes are serial, using step-by-step logic or computations, tend to be invoked deliberately by the agent, and are often associated with a subjective feeling of effort. A person can typically provide a good introspective account of controlled processes. In contrast, automatic processes operate in parallel

Table 1. Two Dimensions of Neural Functioning.

	Cognitive	Affective
Controlled Processes Serial Effortful Evoked deliberately Good introspective access	***I***	***II***
Automatic Processes Parallel Effortless Reflexive No introspective access	***III***	***IV***

Source: Adapted from Camerer et al. (2005, p. 16).

are not accessible to consciousness and are relatively effortless, thus facilitating rapid response and allowing for multitasking, giving the brain amazing power when it comes to certain types of tasks such as visual identification.

The second dimension includes affective and cognitive processes, often seen as emotion or passion, and reason. Although we often associate affect with feeling states, and indeed most affect states do produce feeling states when they reach a threshold level of intensity, Camerer et al. (2005) contend that most affect probably operates below the threshold of conscious awareness. The central feature of affect, therefore, is not the feeling states associated with it, but its role in motivation. As discussed above, all emotions have either a positive or negative valence and many also carry action tendencies, for example, anger and aggression or fear and escape. Therefore, cognition by itself cannot produce action; in order to influence behavior, the cognitive system must operate via the affective system. While it is often presumed that cognition is controlled and affect is automatic, in reality much cognitive processing is automatic as well. Moreover, research on "emotional regulation" reveals many ways that cognition influences emotion and implies a capacity for controlling emotion.

To conclude, emotion is a disposition to action and is part biological and universal and part sociologically and culturally specific. Emotion and cognition evolve together and function together, and neither can be understood without understanding the other. Emotion galvanizes, informs, and motivates cognition, and cognition, in turn, refines, strategizes, and sometimes overrides affective input. As noted above, research suggests that emotion, in the right measure, facilitates, and indeed might be necessary for, rational decision making, assigning priorities to sensory data and sustaining attention, identifying problems and preferences, accelerating strategic reasoning, and helping to store and retrieve memories. In ways only partly understood, emotion animates and helps coordinate problem-solving techniques and their application (Long & Brecke, 2003). Continuing to ignore or marginalize the role of emotions in decision making does not serve us well in our attempts to understand social action and policymaking.

Implications for Policymaking?

Why should this research on the genetic and neurological roots of emotions and decision making have any relevance for the study of political behavior?

Although there are many reasons, two are preeminent. First, it demonstrates that many individual decisions are made automatically without conscious reflection or calculation. "In short, much of what is going on in the human nervous system is occurring outside of conscious awareness" (Gruszczynski et al., 2013, p. 137). Evolutionary psychology offers explanations as to why many human responses occur outside conscious thought. First, it allows for faster and more direct responses and second, keeping information out of conscious awareness might prevent overloading. In the least, this knowledge calls into question the rational decision maker postulated in rational choice models. Moreover, collectively it challenges traditional policy-making institutions by raising questions about group behavior.

Second, it establishes that the affective system is intricately involved in decision making. Contrary to the view that "passion is the enemy of reason," emotions not only have an influence on all decisions but they also have corresponding neurological dynamics that interact with cognitive processes in the brain (Marcus, 2000; McDermott, 2004). Unlike self-interested economic behavior, self-interested political behavior is not necessarily defined by utility functions, and concern about others is not explicitly related to considerations of the welfare of others (Kato et al., 2009).

The common theme that emerges from this research is that many decisions occur at an automatic level until the brain is faced by an unexpected or novel stimulus (see Box 1). When this occurs, the emotional system assists in arousing higher-level conscious systems that are able to analyze and deal with the novelty at hand. In other words, making decisions requires concerted activity across those regions of the brain associated with basic impulses as well as those associated with cognition and reason, although not in any consistent proportion. Thus, many political decisions, especially those that might be perceived as routine, are not necessarily rational, at least not in the sense that prevailing policy-making models assume. For a more detailed discussion of these issues and their implications for democratic theory, see Dustin Tingley who states that the "quest to be able to generate a more complete understanding of political behavior by bridging 'multiple levels of analysis' and connecting political science to neuroscience is exceedingly complicated for a variety of reasons" but necessary if we are to understand political behavior at all levels (2006, p. 6). It would also seem that this knowledge would be invaluable for leaders and political candidates in designing strategies for maximizing political support.

Box 1. Is Political Cognition Like Riding a Bicycle?

Three characteristics of habitual behaviors suggest parallels between political thinking and bike riding. First, both can become routinized and automatic with behavioral repetition. Second, once formed, these behaviors are difficult to explain. Just as it is difficult to describe the coordinated movements that underlie riding a bike, the bases for decision making in many domains become less accessible to conscious inspection over time. And, third, we have imperfect introspective access to the mechanisms that support habitual behaviors and can, therefore, easily lose sight of the forces that trigger and guide their automatic expression. The parallels end, however, when trying to explain "how" we do it. With bike riding we know how inadequate explanations can be, whereas with political thinking we are often unaware of how little insight we actually have in our decision-making processes. Research on introspection has consistently shown that people confidently generate post-hoc narrative accounts of the thinking that supposedly went into a behavior, even when it can be proven to have been driven by factors outside their conscious awareness (Lieberman, Schreiber, & Ochsner, 2003, p. 681).

POLITICAL IDEOLOGY AND POLITICAL ATTITUDES

Although political ideology and attitudes are commonly presumed to have solely environmental causes, recent studies have identified biological influences on an individual's political orientation (Alford, Funk, & Hibbing, 2005). Kanai, Feilden, Firth, and Rees (2011) found that high-level concepts of political attitudes are reflected in the structure of focal regions of the brain. Moreover, brain structure can exhibit systematic relationships with an individual's experiences and skills that can be altered through extensive training and is related to different aspects of conscious perception (Kanai & Rees, 2011). Since understanding how political attitudes change is critical to studying political behavior, stable preferences toward others, such as partisanship and membership in social groups as well as preferences that are more susceptible to change, constitute important subjects for study. Neuropolitical studies have found that the activation of emotion-related areas in the brain is linked to resilient political preferences (Kato et al., 2009). Additionally, the study of ideology and attitudes is one where the interaction of genetic and neural influences is most discernible.

While ideological differences among partisans have been long attributed to social factors, recent research suggests that liberals and conservatives differ on important psychological characteristics, with conservatives demonstrating stronger attitudinal reactions to situations of threat and conflict and liberals tending to seek out novelty and uncertainty (Jost, Glaser, Kruglanski, & Sulloway, 2003). Jost (2006) also found that aversion to ambiguity and a sense of threat and disgust are often traits of conservatives. Moreover, liberal Democrats are more risk accepting than Republicans (Kam & Simas, 2010). Consistent with attitudinal divergence, conservatives have more intense physical reactions to threatening stimuli than liberals (Oxley et al., 2008), while liberals have stronger physiological responses to situations of cognitive conflict (Amodio, Jost, Master, & Yee, 2008). Moreover, Eidelman, Crandall, Goodman, and Blanchar (2012) found that when effortful, deliberate thought is disengaged, endorsement of conservative ideology increases, suggesting that political conservatism might be a process consequence of low-effort thought. Overall, the results suggest that conservatives generally crave closure, prefer to act quickly, and choose instinctive solutions.

There is also mounting evidence that biology influences differences in political attitudes and beliefs with liberals and conservatives displaying dissimilar cognitive styles. A study of young adults revealed that liberals and conservatives present different brain structure, with liberals presenting increased gray matter volume in the anterior cingulate cortex and conservatives in the amygdala (Kanai et al., 2011). Zamboni et al. (2009) used multidimensional scaling and brain imaging to identify which criteria people use to structure complex political beliefs and which brain regions are concomitantly activated and found that three independent dimensions explained the variability of a set of statements expressing political beliefs. Importantly, each dimension was reflected in a distinctive pattern of neural activation: individualism in the medial PFC and temporoparietal junction, conservatism in the dorsolateral PFC, and radicalism in the ventral striatum and posterior cingulate.

Similarly, in their study of differences in brain function between liberals and conservatives who performed a risk-taking task during functional imaging, Schreiber et al. (2013) found that, although the risk-taking behavior of Democrats (liberals) and Republicans (conservatives) did not differ, their brain activity did. Democrats showed significantly greater activity in the left posterior insula, while Republicans showed greater activity in the right amygdala when making a winning risky versus winning safe decision. In fact, a two parameter model of partisanship based on amygdala and insula activations yielded a more accurate model of partisanship than a conventional model based on parental socialization. These results suggest that liberals and conservatives engage different cognitive processes when they think about risk, and they reinforce recent evidence that conservatives show greater sensitivity to threatening stimuli (Schreiber et al., 2013).

Conservatives also have been found to be more organized and persistent in their judgments and approaches to decision making, as indicated by higher average scores on psychological measures of personal need for order, structure, and closure (Amodio et al., 2008). Liberals, by contrast, report higher tolerance of ambiguity and complexity, and greater openness to new experiences on psychological measures (Amodio et al., 2008). One study found that conservatives respond to threatening situations with more aggression than do liberals and are more sensitive to threatening facial expressions (Kanai et al., 2011). This heightened sensitivity to emotional faces suggests that individuals with conservative orientation might exhibit differences in brain structures associated with emotional processing such as the amygdala. In fact, structural MRI data showed increased gray matter volumes in the right amygdala, which was significantly associated with conservatism (Kanai et al., 2011).

Moreover, the results of the Amodio et al. (2008) study were consistent with previous findings that political orientation reflects individual differences in the functioning of a general mechanism related to cognitive control and self-regulation (Alford et al., 2005; Jost et al., 2003). They found that greater liberalism was associated with stronger conflict-related anterior cingulate activity, suggesting greater neurocognitive sensitivity to cues for altering a habitual response pattern. In contrast, stronger conservatism was associated with less neurocognitive sensitivity to response conflicts. At the behavioral level, a liberal orientation was associated with better performance on the response-inhibition task, implying that conservatives would perform better on tasks in which a more fixed response style is optimal.

One study has found that political attitudes are influenced much more heavily by genetics than by parental socialization (Kanai et al., 2011). For the overall index of political conservatism, genetics accounts for approximately half of the variance in ideology, while shared environment including parental influence accounts for only 11 percent. In the case of the variance in people's tendencies to possess political opinions at all, regardless of their ideological direction, genetics is claimed to represent one-third of the variance while shared environment is inconsequential. The expression of genetic influences on political attitudes occurs at the time of early adulthood and persists throughout the adult years. According to Hatemi et al., the "onset of genetic influences at young adulthood raises new questions about the mechanism of genotype-environment interplay and the mechanism underlying the 'impressionable years' model of political socialization" (2009, p. 1153). Their results suggest the parent–child correspondence in attitudes over the long term is more likely to be due to some combination of personal experiences and shared genetic inheritance, not a lasting direct social influence of parents on the attitudes of their offspring.

These controversial findings undeniably have considerable implications for political science since acknowledging a role for heritability in politics alters our understanding of political learning and political cleavages (Alford et al., 2005). There is evidence that inherited attitudes are very different than acquired tastes. They are manifested more quickly, are more resistant to change, and increase the likelihood that people will be attracted to those who share those particular attitudes. Also, to the extent that political ideologies are inherited and not learned, they become more difficult to manipulate (Alford et al., 2005). Conservative parents who try to make their children conservative by controlling their children's surroundings overestimate the importance of these environments. Offspring of such parents are likely to end up being conservative, but less because of the environment created by the parents than the genes passed along by them.

Research on the heritability of political preferences typically has analyzed specific positions on reasonably salient issues on a given culture at a given time. However, as noted in Chapter 2, the possibility that attitudes have a connection to genetics is frequently denounced. A more appropriate response is to think about the nature of genetics, the nature of politics, and the ways the two could be connected, however indirect the connection might be (Beckwith & Morris, 2008). Most observers believe that genes are unlikely to affect attitudes directly. Rather, they affect biological systems that, in turn, affect cognitive processing tendencies that, in turn, affect personality and value traits that, in turn, affect political ideology that, in turn, affect stances on issues of the day. Rather than directly connecting political preferences to genetics, psychologists have been more likely to connect political beliefs to personality, moral foundations, life choices, tastes, and values. Current research on political ideologies suggests that people hold a political preference if they have a consistent view on issues of the day, if their stance on one issue constrains their stances on other issues, and if they are aware of the meaning of commonly employed ideological labels.

In order to be considered ideological, an individual would need to understand what it means to be liberal or conservative and to have a collection of political beliefs that is consistent and that logically fits within one of the labels (Beckwith & Morris, 2008).

In direct contrast to the arguments above, however, some contend that whatever constitutes ideology, its ultimate cause is environmental, and any particular ideology can only make sense within the context of a unique political culture. Smith, Oxley, Hibbing, Alford, & Hibbing (2011a, 2011b) argue that the dominant view within political science that ideology is optional, and, in fact, largely absent in most people, and merely a collection of issue positions not reliant on deeper, more universal, psychological tendencies, has resulted in a shallow and narrowly political conceptualization of the term and hampered the ability of political scientists to integrate their findings with those psychologists working on the same topic. Thus, recent developments have seen a tendency of moving away from the traditional definition of ideology. If political ideology is to become useful and meaningful, it is necessary to accept that ideology is not a superficial label or a bundle of topical positions but rather a central component of an individual's general life orientations. In the past, political scientists have measured ideology in ways that guaranteed it could only be seen as emanating from particular cultures, and, according to Smith et al. (2011a, 2011b), this unfortunate vision of ideology should end.

A four-nation study evaluated a wide variety of ideological measures and found significant mean differences across populations, measures, and time periods, although variances were comparable and heritabilities were consistently confirmed in the 0.30–0.64 range (Hatemi et al., 2014). Moreover, while genetic influences could be statistically equated across populations, environmental components could not, suggesting that the relative importance of genetic influences is common across cultures, whereas the influence of family and personal environments varies greatly in explaining the variance in attitudes. Genetic influences accounted for an even greater proportion of individual differences in models which evaluated ideological positions of the entire family, while direct learning from parents accounted for a minimal portion of the variance on political attitudes (Hatemi et al., 2010). Utilizing data from the Minnesota Twin Family Study, Cranmer and Dawes (2012) show that a moderate portion of individual differences in the degree to which one's foreign policy preferences are hawkish or dovish can be attributed to a genetic source shared with political ideology, thereby confirming that ideology could be the causal pathway through which genes affect policy preferences.

Although a person's general ideology can be understood as extending into every facet of life, political ideology is the political manifestation of these deeper inclinations toward a variety of features of our existence, not simply a superficial and arbitrary summation and labeling of issues and attitudes. Just as personality traits and values flow from a combination of the environment and cognitive/biological tendencies, people's broad political preferences flow from their values and personality traits (Smith et al., 2011a, 2011b). Acceptance that there are biological and even genetic precursors to specific political attitudes and behaviors is a hard sell because the traditional view of politics places great weight on smaller issues arising within a narrow cultural context, thereby making it much more difficult to imagine how broad biological forces could be at work (see Box 2).

Overall, Alford et al. (2005) estimate that genetic factors account for about 32 percent of a person's political ideology. Moreover, while a model of partisanship based on mother's and father's party identification predicts about 69.5 percent of self-reported party choices, the utilization of activations in the amygdala and the insular cortex correctly predicts

Box 2. Predisposed.

Buried in many people and operating largely outside the realm of conscious thought are forces inclining us toward liberal or conservative political persuasions. Our biology, not always reason and the careful consideration of facts, predisposes us to see and understand the world in dissimilar ways. Moreover, these predispositions are responsible for a significant portion of the political and ideological conflict that marks human history. There is overwhelming evidence that people differ politically not just because they grew up in different cultures or were presented with different information. The universal rift between conservatives and liberals endures because people have diverse psychological, physiological, and genetic traits that influence much of what makes people who they are, including their orientations to politics. The key to getting along politically, then, is not the ability of one side to persuade the other side to see the error of its ways but rather the ability of each side to see that the other is different, not just politically, but *physically* (Hibbing, Smith, & Alford, 2013).

82.9 percent of the observed choices of party, therefore significantly outperforming the prevailing parental model and offering more powerful insights than the traditional tools used by political science (Schreiber et al., 2013). Although past studies have shown genetic variation to contribute to variation in political ideology (Alford et al., 2005) and strength of partisanship (Settle, Dawes, & Fowler, 2009), the portion of variance in political affiliation explained by activity in the amygdala and the insula is significantly larger, thus suggesting that acting as a partisan may, in fact, alter the brain beyond the effect of heredity. The interplay of genetic, epigenetic, and environmental effects may also be driving the observed correlations between the size of brain regions and political affiliation (Kanai et al., 2011). As cogently summarized by Hatemi and colleagues:

> Genetic influences on political ideology are not boundless and social influences are far from irrelevant to the transmission of ideology. Political ideologies are complex, interactive, and environmentally contingent and phenotypic heterogeneity is undoubtedly present .... The formulation of an integrated theory of political ideology requires the integration of genes and environment, embedded within a developmental framework, that allows for the incorporation of parental investment, social groups, cognition, perception, education, and a wide array of neurobiological mechanisms at the very least. In such a framework, there is scope for the roles of genetic expression and the social environment to change during development in ways that are consistent with the acquisition and application of social values and behavior .... In this way, we use models of genetic variation to provide the critical bridge that merges the research in the social and life sciences into a more comprehensive model regarding the nature of ideology. (2014, p. 21)

PERSONALITY AND POLITICAL ATTITUDES

Although, as discussed in Chapter 2, personality studies have a venerable history in political science, overall political behavior scholars have paid scant attention to the biological groundings of personality, instead focusing on social variables such as socioeconomic status, party identification, and short-term forces. At best, heredity has been considered a negligible factor. In the past decade, however, a cohort of political psychologists employing data from behavioral genetics has begun to explore the connections between political attitudes and behavior and personality traits rooted in biology. Unsurprisingly, this emergent research on personality and political behavior has encountered considerable skepticism (Hatemi & McDermott, 2012a, 2012b).

Importantly, psychologists have established that genes and other biological factors "account for most of the variance in personality traits" which remain stable throughout life. In turn, these traits influence our political activity (Mondak, Hibbing, Canache, Seligson, & Anderson, 2010, p. 89). The means of numerous estimates of the heritability of the Big Five personality traits (see Chapter 4) indicate that extraversion has a heritability coefficient of about 0.53, agreeableness 0.42, conscientiousness 0.46, emotional stability or neuroticism 0.49, and openness or curiosity 0.54 (Alford & Hibbing, 2008). Moreover, personality traits appear to be relatively permanent and directly involve behaviors.

Although these studies predominantly define personality by the Big Five traits, some focus on specific traits such as altruism, shyness, efficacy, and conflict avoidance (Blais & Labbe-St Vincent, 2011) or aggressive (Kalmoe, 2013). Cumulatively, these works have identified links between personality trait dimensions and numerous variables, including ideology, partisanship, policy attitudes, political interest, efficacy, knowledge, information exposure, participation, and group identification and favoritism to in-group members (Weber, Johnson, & Arceneaux, 2011). Hibbing, Ritchie, and Anderson (2011) found that personality traits consistently influence the nature of political discussion: extraversion was linked with more frequent discussion, emotional stability with increased conversations with people of differing viewpoints, conscientiousness with more discussion of local issues, and openness to experience with the likelihood of being influenced by the people with whom they discuss politics. They conclude that the "question is no longer whether personality matters but how exactly it matters" (Hibbing et al., 2011, p. 602).

Dietrich, Lasley, Mondak, Remmel, and Turner (2012) expanded this analysis to legislative behavior and concluded that variance in personality influences legislators' attitudes regarding important facets of their careers. Moreover, the personality traits that matter for political ambition – extraversion and emotional stability – differ from the one that corresponds with the greatest satisfaction in legislative office – openness to experience. Similarly, Kalmoe (2013) found that trait-aggressive individuals tend to be more confident in the effectiveness of aggressive responses to problems and to hold attitudes and values characterizing aggression as a positive response. They are also more likely to anticipate aggressive acts from others and to perceive ambiguous actions by others as hostile. Consistent with an interactive model, these differences are attributable to long-term social learning processes that interact with trait aggression and reinforce it over time.

In their path-breaking study, Mondak et al. (2010) argue that certain types of political participation can be predicted by interacting personality traits with environmental variables and that personality research will eventually supplant genetic research. Whereas Mondak et al. use personality traits to predict participation, Gerber, Huber, Doherty, Dowling, & Ha (2010) argue that personality traits influence our ideology. Individual personality is shaped by experience, including family dynamics, cultural forces, work, and educational experiences, but compelling evidence shows that highly stable personality differences affect how individuals respond to the environmental stimuli they encounter. Thus, personality traits can be viewed as predating, rather than being caused by, social and political influences, offering an opportunity to examine how elemental, enduring personality differences affect an array of social outcomes. Thus, Gerber et al. hypothesize that four of the Big Five traits influence ideology. The only exception is extroversion, which they expect to influence political participation (as Mondak et al., 2010, demonstrate), but not ideology. The authors confirm all of their predictions except that emotional stability would cause a person to lean left on social issues. As it turns out, emotionally stable individuals lean right while those with neurotic traits lean left. Thus, conservatives come across as hard-working, organized, closed-minded, and emotionally stable, while liberals are lazy, disorganized, open-minded, and neurotic. Gerber et al. (2010) conclude that the effects of personality rival the effects of education and income, although the effects can be contextual, in particular regarding race.

In a later article, Gerber, Huber, Doherty, Dowling, and Panagopoulos (2013) found that Big Five personality traits affect how people respond to the costs and benefits of voting following get-out-the-vote appeals. One trait in particular, openness, is associated with broad persuadability, while others shape responses to particular types of messages. In some cases, the conditioning effects of Big Five traits are substantial. Likewise, in a German study, Schoen and Steinbrecher (2013) demonstrate that agreeableness and emotional stability increase electoral participation but that attitudinal variables, including party identification, civic duty, political interest, and efficacy act as intervening variables that mediate the impact of personality on turnout. Overall, the evidence suggests that openness, agreeableness, and extraversion render attitudes somewhat less powerful in affecting turnout, while conscientiousness and emotional stability increase the impact of certain attitudes. Moreover, indirect and conditioning effects shape attitudes in a way conducive to higher turnout and make them more powerful in affecting voter participation.

Similarly, Blais and Labbe-St Vincent (2011) found that political interest is more strongly influenced by personality than by sense of civic duty, while altruism and shyness are associated with both interest and duty. In contrast, personal efficacy is correlated with duty, but not with political interest. Those who feel in command of their life may be more inclined to believe that voting is one of the "good" things that one should do. If we can agree that, given the low probability of casting a decisive vote, the rational choice for an egocentric citizen who wishes to maximize his or her personal utility is to abstain, it makes sense to suppose that the propensity to vote is related to how egocentric or altruistic one is. They conclude:

> These findings suggest that personality traits do matter. They do affect political interest and sense of civic duty as well as the propensity to vote. The most consistent findings concern altruism, which is positively associated with each dependent variable. Our assumption is that personality traits develop earlier in life and tend to be more stable than attitudes toward politics, which are formed somewhat later as people become more exposed to political affairs. As a consequence, the impact of personality on the likelihood of voting is basically indirect-that is, personality contributes … to nourishing attitudes such as interest and duty, which themselves more directly affect the decision to vote or not to vote. (Blais & Labbe-St Vincent, 2011, p. 407)

This view is consistent with the claim of Gerber, Huber, Raso, and Ha that "personality can be viewed as something that predates, rather than is caused by, social and political influences" (2009, p. 3). From such a perspective, personality traits can be construed as primary distant variables which, in combination with other environmental factors, shape the proximal political attitudes upon which political science research typically focuses (Blais & Labbe-St Vincent, 2011).

As presented in Chapter 2, based on such findings, Mondak et al. (2010) proposed an integrative framework for the study of trait effects on political behavior (see Figure 2 of Chapter 2). People's enduring psychological tendencies are reflected in their traits, but one cannot understand the effects of personality without accounting for the environment, just as one cannot understand the effects of the environment without accounting for personality. Neither environmental factors nor personality traits have much of a direct effect on political behavior. Instead, most of the effect is interactive. For example, if a person has an extroverted personality type, and if a form of political participation is social like a caucus as opposed to casting an anonymous vote, then one would expect that person to participate.

In multiple articles, however, Verhulst and various coauthors reject a causal link between personality traits and political ideology (e.g., see Verhulst, Eaves, & Hatemi, 2012; Verhulst, Hatemi, & Martin, 2010).

When they decomposed the variance between personality traits and ideologies, they found that a sizeable majority of the covariance between personality traits and attitudes is a function of latent shared genetic influences. Moreover, the majority of genetic variance within attitudes was not explained by the genetic variance on personality traits. Thus, the relationship between personality traits and political attitudes is not causal in nature, but rather the result of an underlying genetic variable that is a common cause of both personality traits and political attitudes. Therefore, the assumption that personality traits developed in infancy *cause* the subsequent development of political attitudes in adulthood is incorrect. Recent scholarship has demonstrated that political attitudes develop much earlier than previously suspected, the precursors of which are present prior to a child's first year in school and are influenced by genetic factors (Hatemi et al., 2010). The foundational elements of political attitudes exist as part of our core disposition and appear to be as important in shaping our behavior as our personalities (Verhulst et al., 2010).

Gallego and Oberski (2012) contend that the effect of personality on political participation is mediated by classical predictors, such as political interest, internal efficacy, political discussion, and voting as a civic duty. Their study confirms that the effects of personality traits on voter turnout and protest participation are sizeable but indirect and mediated by attitudinal predictors. Personality affects participation indirectly because it shapes the propensity to acquire attitudes and behaviors which in turn are predictors of participation. Despite being indirect, however, the effects of personality traits on political participation are meaningful.

> It is plausible to conclude from our findings that there is a multiple-step funnel of causality in which personality, socio-demographics and possibly other factors influence political attitudes, which in turn are the most proximate causes that affect political behavior. Such a funnel of causality approach can be adopted to study the impact of other predictors of participation such as genetics or, by including interactive effects, contextual features. (Gallego & Oberski, 2012, p. 444)

Personality is, therefore, integral to understanding why some people participate in politics while others do not, as well as ideology.

POLITICAL GROUNDINGS

Politics is more than attitudes and voting; efficaciousness, political sophistication, political engagement, and other forms of participation are also important. Although research on the impact of genetics in most of these areas is embryonic, it is growing (Hatemi & McDermott, 2012b). As discussed earlier in this chapter, the foundational elements of political behavior, such as trust, empathy, and prosociality, are well established in social neuroscience and behavioral genetics. For example, twin studies have revealed the importance of genes for prosocial behaviors. Moreover, heritability ranging from 0.40 to 0.70 has been observed for self-reported measures of altruism, cooperativeness, trust, and nurturance (Dawes et al., 2012). Similarly, twin studies on populations in Sweden, Australia, Denmark, and the United States have uncovered small to large genetic influences on social trust, political efficacy, political sophistication, sense of citizen duty, and political interest, many of which share a common genetic factor to voter turnout and participation (Klemmensen et al., 2012; Littvay, Weith, & Dawes, 2011; Oskarsson, Dawes, Johannesson, & Magnusson, 2012; Sturgis et al., 2010).

Politically sophisticated individuals make sense of politics in terms of an overarching ideology, while those without it must rely on cruder heuristics, such as partisan and group-based identities, to guide their political attitudes (Arceneaux, Johnson, & Maes, 2012). Therefore, political sophisticates who reason about politics using abstract principles closely approximate an ideal of democratic citizenship. In the past, political sophistication has been measured using various indicators: educational attainment, interest in politics, knowledge of basic political facts, as well as the degree to which attitudes on various political issues are consistent with the liberal-conservative ideological continuum as constructed by elites (Weisberg & Nawara, 2010). However, the SSSM assumption that politically sophisticated parents teach their children to be active, interested, and sophisticated citizens like themselves conflicts with the extant evidence flowing from biological-based research. Schreiber's (2005) work utilizing fMRIs of brain activity during political cognition provides confirmation that differences between political sophisticates and political novices arise from their use of different neural substrates. Moreover, Arceneaux and colleagues show that the behavioral and attitudinal facets of political sophistication are mostly explained by variation in genes and the unshared environment. They conclude:

> Virtually none of the variance in education, political interest, political knowledge, and ideological constraint is explained by the common environment. These findings are in stark contrast to the SSSM's emphasis on family socialization as a source of political sophistication and corroborate previous research, which shows that each of these factors account for overlapping but distinct aspects of political sophistication. (2012, p. 11)

They also found that the bivariate correlations among education, political interest, and political knowledge are largely accounted for by genetic factors, while the correlations between ideological constraints and both interest and knowledge are largely a function of the unique environment. To some degree, education, political interest, political knowledge, and political sophistication derive from a shared source – be it a common set of genes or a common set of environmental factors experienced in the unique environment.

Recent studies have also found that people who are risk accepting participate in political life more because politics offers novelty and excitement. Kam (2012) found a significant relationship between risk attitudes and general political participation, but one that varies with the type of political activity: people who are more risk accepting are more likely to participate in general political acts, but no more or less likely to turn out in elections. In their twin study, Littvay et al. (2011) show that many measures of efficacy are highly heritable and that their covariance with voter turnout predominantly can be attributed to genetic factors. Moreover, Klemmensen et al. (2012) in twin studies from Denmark and the United States conclude that the heritability of political participation and political efficacy is remarkably similar across cultures and that most of the covariation between efficacy and participation is accounted for by a common underlying genetic component.

Working under the theory that political participation is some function of prosociality, twin and kinship studies have revealed that genetic influences account for around 0.53 of the variation in voter turnout (Fowler, Baker, & Dawes, 2008). Country studies that measure participation in different ways, such as donating money, writing to a member of parliament, attending rallies, or volunteering to work on campaigns have reported similar results (Klemmensen et al., 2012). As noted in Chapter 2, genetic studies have revealed that variants of dopamine (D2, DRD4) and serotonin (5HTT) genes influenced voter turnout and general political participation (Fowler & Dawes, 2008).

VOTING BEHAVIOR

Voting is considered as one of the most important decisions that individuals can make in a democratic society and, thus, it is no surprise that the factors influencing voting turnout and choice have long been of interest to political researchers and have spawned many models (see Chapter 2).

As seen earlier, however, conventional approaches to the processes through which voters come to support particular political candidates tend to ignore the effect of emotions and instead view voters as responsive to the candidates' position on the issues, party affiliation, and physical traits such as race, gender, and age. Although these factors undeniably influence voter choice to varying degrees, neglecting the influence of emotion discounts the evidence summarized in this chapter (see Box 3). Recently, brain imaging studies of the neural bases of partisan political judgment and candidate preference have provided some of the most innovative and controversial research on political behavior. Overall, this research strongly suggests that neural substrates and biological mechanisms shape our preferences and decisions regarding candidates (Friend & Thayer, 2011).

By focusing chiefly on environmental influences on vote choice with no explicit role for biological factors, traditional voting choice theories make critical assumptions that differ from typical assumptions made by behavioral geneticists and neuroscientists (Hatemi, Medland, Morley, Heath, & Martin, 2007). To remedy this, over the past few years claims that establish a link between genes and political behavior have been enforced by twin

Box 3. One Finds What One Looks For.

Much of modern political science assumes that voters consciously select candidates based on their estimates of how the candidate will satisfy their values and material needs. This is a controlled and cognitive process, and more likely to activate an executive region of the brain such as the lateral prefrontal cortex that implements the strategic pursuit of goals. Quite plausibly, however, voters might also prefer candidates in part because they are physically attractive, which would correspond to automatic/affective processing because it elicits a noncontrolled response that is highly affective and likely processed in part by elements of the limbic portion of the brain. Explanation of voter behavior thus might have two different theories and two different predictions of neurological activity (Todorov et al., 2005). "Of course, elements of both processes occur. The point is twofold: first, the type of political explanation being offered affects the type of neurological evidence that would be suggested; second, if both processes occur, fMRI could be used to study their relationship" (Tingley, 2006, pp. 11–12).

research and led to the creation of genopolitics (Joseph, 2010). As detailed in Chapter 3, the classical twin method compares the trait resemblance of reared-together MZ (monozygotic, identical) twin pairs with the resemblance of reared-together same-sex DZ (dizygotic, fraternal) twin pairs. Twin researchers and even most of their critics agree that MZ twins resemble each other more than same-sex DZ pairs for most behavioral and psychological traits (Joseph, 2010).

As discussed in Chapter 2, second-generation biopolitics scholars influenced by behavioral genetics argue that findings from twin studies offer solid evidence that genetic factors play an important role in politics. Thus, political behavior joins a long list of other intuitively non-genetic behaviors now claimed for genetics on the basis of twin research, including anorexia, loneliness, perfectionism, and religiousness. Vocal critics, however, have argued that heritability claims based on twin studies generally cannot be trusted due to the confounding influence of the greater environmental similarity of MZ twins. According to Joseph (2010), twin studies of political attitudes and behavior provide no scientifically acceptable evidence in support of genetic factors. "Moreover, recent gene-finding claims use the same methodology that has led to 30 years of false-positive findings for behavioral traits and psychiatric disorders, and there is every reason to believe that their claims will also suffer the same fate" (Joseph, 2010, p. 213). Therefore, while there seems to be considerable research supporting the view that political behavior is influenced by genes, widespread disparagement of the methods and conclusions persists.

Although the choice to vote for a particular candidate or party is unlikely to be heritable, it remains an open question whether or not the act of voting is heritable. As Fowler and associates state:

> It is extremely unlikely that such efforts will uncover a voting gene, however there is some suggestion that there is some set of genes who expression – in combination with environmental factors – regulates political participation. The task of finding out which genes they are and what physical function they have will improve our understanding of the biological processes that underlie these complex social behaviors and may also shed light on their evolutionary origin. (2008, p. 244)

The realization that participation is heritable has already helped to generate additional theories of turnout, partisanship, and prosocial behavior. According to Settle et al., genes are the institutions of the human body, and we ignore them at our peril if we want to develop a full understanding of human psychology and political behavior (2009).

While traditional research on voting behavior has always presumed that a candidate's appearance can affect voter choice, recent cognitive neuroscience studies provide us with a better understanding of how emotion and motivated reasoning influence political judgment and voter choice. Behavioral studies have shown that appraisals about candidates' physical appearance do correlate with election outcomes (Ballew & Todorov, 2007; Todorov, Mandisodza, Goren, & Hall, 2005), suggesting that information derived from visual appearance alone (so-called "thin-slice" information) affects voter choice. Much of the research on how appearance affects voting focuses on the facial expressions of candidates and on what regions of the brain respond to the emotional displays. This work has confirmed the importance of physical attractiveness and revealed a very strong role for "facial dominance," in voter decisions. Considerable research also has shown that the attractiveness of the human face is highly correlated with the degree of bilateral symmetry of facial features (Tingley, 2007, p. 154) and that perceptions of physical attractiveness are not highly culture specific. Thus, the culturally relative portrayal of attractiveness suggested by the SSSM model is empirically incorrect or, at least, overstated.

Moreover, these findings suggest that evolved preferences strongly influence political and economic decisions; visual cues, for example, demonstrably complement beliefs about a candidate's viability. Despite the importance of elections, research shows that electoral outcomes can be predicted on the basis of snap judgments of the faces of the candidates (Chiao, Iidaka, & Gordon, 2008). One study found that judgments of competence based on seeing American political candidates' faces for only 100 milliseconds accurately predicted 59.6 percent of the electoral outcomes (Ballew & Todorov, 2007). Similarly, Hehman, Carpinella, Johnson, Leitner, and Freeman (2014) found that gender categorization dynamics during the initial milliseconds after viewing a female politician's face are predictive of her electoral success, especially in more conservative areas. These findings that people automatically make judgments from facial appearance were replicated by Todorov, Pakrashi, and Oosterhof who note: "We showed that trustworthiness judgments are made within a single glance of a face" (2009, p. 830). Moreover, brain imaging techniques, in combination with behavioral studies on the psychological mechanisms involved with appearance-based judgments, have begun to offer valuable insights into the underlying neural processes of voter preference.

A fundamental question in politics is the extent to which voters' decisions are driven by positive motives, which induce them to vote for a candidate who they like, or by negative ones, which bring them to, in effect, vote against the candidate that they most dislike (i.e., negative voting). To find out if the effect of appearance on voting is more strongly influenced by positive or negative emotions, Spezio et al. (2008) conducted brain imaging studies in which two separate groups of participants viewed pairs of unfamiliar real politicians (both Republican and Democrat) from previous elections. In the first study, images of losing candidates elicited greater activation in the insula and the ventral anterior cingulate than images of winning candidates, suggesting that negative attributions from appearance exert greater influence on voting than do positive. In a second study, when negative attribution processing was enhanced, images of losing candidates again elicited greater activation in the same brain regions. The authors conclude that these findings support the view that negative emotions play a critical role in mediating the effects of appearance on voter decisions, an effect that may be of heightened importance when other information is absent (Spezio et al., 2008). In other words, these "thin-slice" attributions play the largest role in influencing voters who have little knowledge of the candidates, thereby reinforcing already strong evidence that when voters know little about a candidate, perceived negative aspects of a candidate exert a stronger influence on voter turnout than positive aspects (Stevens, Sullivan, Allen, & Alger, 2008). Negative campaigning can pay off, especially at the margins.

In another fMRI study, Kato et al. (2009) imaged subjects after viewing negative campaign videos and found those with stronger activation in the dorsolateral PFC lowered their ratings for the candidate they originally supported more than did those with weaker fMRI activation in the same cortical area. Moreover, subjects showing stronger activation in the medial PFC tended to increase their ratings for the candidates attacked in the negative campaign videos more than did those with weaker activation. Interestingly, the same regions were not activated while viewing negative advertisements for cola, which were used for purposes of comparison. These results imply that neural activity after exposure to negative information about previously supported political candidates was linked to cognitive control of socially relevant stimuli. The activation of distinct prefrontal areas indicates that different kinds of cognitive controls were associated with opposite responses to negative information about the previously supported candidates, that is, they were associated with increasing and decreasing political support (Kato et al., 2009).

Although these studies have confirmed that a candidate's appearance can induce both positive and negative affect, it remains unclear what attributes of a person's appearance signal negative traits that could turn off voters and result in election loss. A cross-cultural fMRI study of the influence of the amygdala on political judgments and voting decisions may hold some insight into the involvement of "political intangibles" (Rule et al., 2010). American and Japanese university students viewed and made judgments on facial images of candidates from the 2004 and 2006 U.S. Senate elections and from the 2000 election of the Japanese Diet, respectively. The study found that with both sets of participants, "the bilateral amygdala was significantly more responsive to the faces of politicians for whom participants chose to vote" and "participants showed a greater overall amygdala response to the faces of out-group candidates than they did the faces of in-group candidates" (Rule et al., 2010, pp. 353–354). The authors also uncovered increased amygdala activation in subjects when they viewed candidates for whom they voted in a simulated electoral task. Likewise, Kaplan, Freedman, and Iacoboni (2007) examined neural activation patterns among partisans viewing images of the 2004 presidential candidates and conclude that compared with viewing one's own candidate, viewing a candidate of the opposing political party produced signal changes in cognitive control circuitry in the dorsolateral PFC and the anterior cingulate and in emotional regions such as the insula and anterior temporal poles.

Overall, these findings suggest that the amygdala may be important in evaluating candidates for political office and that the neural basis for electoral decisions extends across cultures. One explanation for stronger amygdala response to the faces of cultural out-group candidates could be that the amygdala is highly responsive to novel stimuli such as the faces of out-group members (Van Bavel, Packer, & Cunningham, 2008). Furthermore, because these findings are consistent with studies of the neural underpinnings of prejudice and bias involved in evaluating out-group members, it appears that voter judgment and decision making are not only influenced by rational cost–benefit evaluations of a candidate's issues and policies, but are also, and possibly more so, the product of virtually automatic emotional reactions to adaptive biological mechanisms associated with maintaining in-group cohesion and survival (Friend & Thayer, 2011).

Along similar lines, Westen (2007) argues that a neuronal approach to emotion-biased motivated reasoning provides a more accurate understanding of the formation of partisan political judgment and voter choice. He contends that the prevailing theory that the voter's mind is a cool calculator that makes decisions by weighing the evidence bears no relation to how the brain actually works. In politics, when reason and emotion collide, emotion invariably wins: citizens' values and feelings usually outweigh their interests and beliefs at the polls. Neglecting the emotions for the sake of reason, therefore, is not only factually misleading, but can also be politically devastating because emotions frame how people think about the world and provide the motivation necessary to do something about those beliefs. Evolutionarily important goods like survival, reproduction, kin, and sociality are likely to serve as emotional hot-buttons that can be manipulated in policy debates to activate neural networks that motivate desired responses. For Westen, elections are decided in a marketplace of emotions filled with values, images, analogies, moral sentiments, and moving oratory, in which logic plays only a supporting role. He shows, through the evolution of the passionate brain and a summary of 50 years of American presidential and national elections, why campaigns succeed and fail, and argues that the evidence is overwhelming that three things determine how people vote. In order, they are: their feelings toward the parties and their principles, their feelings toward the candidates, and, if they haven't decided by then, their feelings toward the candidates' policy positions. Although one cannot change the structure of the brain, one can change the way you appeal to it.

In an earlier article, Westen, Blagov, Harenski, Kilts, and Hamann (2006) reported the results of their fMRI study which showed that self-described Democrats and Republicans responded to negative remarks about their political candidate of choice in systematically biased ways. Specifically, when Republican test subjects were shown self-contradictory quotes by George W. Bush and when Democratic test subjects were shown self-contradictory quotes by John Kerry, both groups tended to explain away the apparent contradictions in a manner biased to favor their candidate of choice. Similarly, none of the circuits involved in conscious reasoning were markedly engaged, while areas of the brain controlling emotions, particularly the amygdala and/or the cingulate gyrus, showed increased activity as compared to the subject's responses to politically neutral statements associated with politically neutral people. In addition, the dorsolateral PFC and anterior cingulate cortex are activated when subjects viewed images of a candidate of a political party different from their own (Kaplan et al., 2007). Furthermore, when subjects were presented with information that exonerated their candidate of choice, areas of the brain involved in reward processing (the OFC and/or striatum/nucleus accumbens) displayed increased activity. Accordingly, "... essentially, it appears as if partisans twirl the cognitive kaleidoscope until they get the conclusions they want ... emotionally biased judgments when they have a vested interest in how to interpret 'the facts'" (Westen et al., 2006).

In a related area of investigation about elections some researchers have examined the hormonal effect of winning and losing for the voters. For instance, in their study of the 2008 presidential election, Stanton et al. (2009) found that vicarious victory and defeat via democratic elections have the same effect on men's testosterone levels as does winning or losing an interpersonal dominance contest. They found that men who voted for the winner Obama had stable post-outcome levels of testosterone, and men who voted for losers McCain or Barr had decrements in their testosterone levels. In other words, male voters experienced the outcome of winning or losing vicariously through their candidate even though they did not personally win or lose the election.

CONCLUSIONS: BIOLOGY AND POLITICAL BEHAVIOR

Although much of it remains highly controversial and often speculative, there is considerable evidence from the life sciences that human behavior is influenced and shaped by evolutionary and biological factors. Ironically, while the contribution of genetic and neurological components to political behavior is largely recognized by those who study human behavior in other disciplines, until recently political scientists have failed to include biological variables in their efforts to explain the political behavior among either the masses or the leaders. Moreover, despite strong support for inclusion of physiological variables such as general health, physical fitness, nutrition, stress, illness, disability, and aging in an interactive model of political behavior, they have been all but ignored by mainstream political science. With few exceptions, textbooks on political behavior have not taken a life sciences approach, and most fail to even mention potential biological determinants of political behavior and continue to treat the political person as an empty organism influenced by socializing agents throughout life.

The wide-ranging literature summarized here, although often exploratory and provisional, demonstrate that, with some initiative and creative thought, biological variables can fruitfully be incorporated into the study of the political behavior. Moreover, when they are included, they appear to be potentially valuable independent variables. As noted earlier, the tendency to summarily dismiss such variables from analysis is isolating political science from the contemporary study of political behavior, which increasingly is becoming the domain of behavioral geneticists, neuroscientists, psychologists, and a handful of political psychologists. As a result, much of what has been offered as explanations of political behavior by conventional political science is dubious to persons with knowledge of biology and behavior.

CHAPTER 6

BIOPOLICY: SOCIAL ISSUES

Odelia Funke

Biopolicy refers to an approach that looks to the life sciences and concepts from evolutionary biology to explain important aspects of social conditions and interactions. Bringing insights about natural systems into political and social analysis requires a multidisciplinary approach. This means moving outside of disciplinary silos to cultivate a broader understanding, from evolutionary biology and the findings of the biological sciences. A biopolitical perspective can enrich explanations of social and political issues and delineate significant considerations for analyzing alternative policy strategies.

Humans are part of the Earth's natural systems, inextricably tied to nature's interconnected networks and feedback loops. Evolutionary theory draws attention to the pervasive interaction between humans and nature. It provides a wealth of insights about the context for life in a dynamic equilibrium within which natural systems survive and flourish, or risk annihilation. As we have argued here, for social analysis to ignore this mutual interaction is perilous as humans are a part of nature, not separate from it. Humans respond to, but also define and change, biosocial and physical environments. With increasingly sophisticated technologies, humans have acquired the ability to create significant ripples in the biosphere, changing the natural systems upon which life depends, and thus influencing evolution itself. However, nature is not merely an engineering problem to be solved or an enemy to be conquered (common post-Enlightenment attitudes), but rather a network of systems upon which human life depends. Human

Politics and the Life Sciences: The State of the Discipline
Research in Biopolitics, Volume 12, 179–204

ISSN: 2042-9940/doi:10.1108/S2042-994020140000012005

activities that disrupt or damage natural processes ultimately affect human life. As discussed in detail in Chapter 7, climate scientists present convincing evidence that climate change is underway, and that it is significantly affected by human activities. As a result, societies will be forced to adapt and respond to many new challenges in order to survive. Humans are slowly relearning what ancient societies knew, that we should seek to live in harmony with nature. If we do not attempt to understand the interplay and limits of natural systems and adjust policies accordingly, the consequences could be devastation of ecosystems and societies (Diamond, 2005; Kolbert, 2014).

As human activities threaten to disrupt natural systems, interdisciplinary research will better enable social scientists to weigh the implications of policy choices such as those that entail long-term damage or change to natural systems as well as the potential for seeking cooperative solutions among groups. The goal of the social scientist, then, must be a deeper, more accurate understanding of the nature of human society.

Evolutionary and anthropological research studies the human instincts and drives that underlie social interactions. Science confirms that humans are by nature social, an observation with far-reaching implications. The principal mode of human existence has been in social communion with others, not as isolated, egocentric actors. Community life, mutual cooperation, and empathy are fundamental to the human animal (see Chapter 5). Sociability is deeply ingrained as it is part of primate behavior. Moreover, cooperation to promote the common good is arguably more powerful and prevalent in human society than conflict (de Waal, 2005, 2006). Evolutionary success for humans and other primates has depended more upon social bonds of cooperation than selfish individual pursuits, more than war and conflict (Corning, 1983). In addition to recognizing the importance of sociability, a life sciences perspective emphasizes holistic and systemic models for analysis. To understand societal problems and possible solutions requires that one think in terms of interconnected, living, adaptive systems, not rigid mechanistic models. Interactions among humans and natural systems can have consequences far into the future; therefore, societal issues by necessity need to be framed and addressed with an eye to intergenerational effects.

Biopolitics, then, offers a corrective to contemporary social science by providing valuable insights into public policy issues dependent on life sciences knowledge, addressing issues in areas such as public health, medicine, agriculture, environment, energy, education, populations and aging, gender and race, body/brain functions, social psychology, leadership, biotechnology, and nanotechnology. Other policy areas, such as transportation and housing, could also benefit from improved understanding of human social response mechanisms. This chapter advocates a biopolitical perspective to better identify the factors underlying pressing societal issues and feasible policy solutions. Examples of the obesity problem and public health, and a more detailed analysis of energy policy and hydraulic fracturing, are used to demonstrate this need.

SHORTCOMINGS OF CURRENT SOCIAL POLICY ANALYSIS

Since the Enlightenment, Western social science has emphasized a distinction between humans and the natural world by focusing on "man the maker," which celebrates humankind's ability to dominate other life forms and materials. Technology's success in fueling continued expansion of human dominance over nature (deemed progress) lies in the mechanistic models of the physical sciences. The key to control has been a continually expanding knowledge and use of scientific methods to shape and reshape materials to suit human needs and desires. Hoping to emulate the demonstrated success of modern science, social scientists have embraced models from mathematics and the physical sciences and largely rejected the life sciences as a core element for analyzing political issues and policy alternatives. We argue, however, that living systems provide better models for social life and that research on evolved human characteristics and needs should inform our conceptual models for society.

As discussed in the earlier chapters, rational utilitarian choice models, widely used for public policy, are based on a hypothetical construct of the "rational man" who weighs options by calculating utilitarian benefits compared to costs; risks being part of the potential cost. This approach conceives of human choice with full knowledge of options, disentangled from prejudice, ignorance, emotion, modern marketing techniques, and deception, and similarly separated from human communal experience or ties of love, friendship, and kinship to influence the "rational" choice. As is clear from the evidence presented in Chapter 5, however, such models do not start with a defensible or realistic model of human nature and reflect an essentially distorted view of social actors. They reify complex issues into monetized costs and benefit numbers to compare, as if the biosocial and physical environments could be sufficiently represented in this way.

In contrast, biopolicy starts, as did the ancient Greek philosophers, with humans as social animals (e.g., Dobzhansky & Montagu, 1962). Human choice does not occur outside of a cultural and social context, and in fact, the social context strongly influences thought and action. Furthermore, humans and ecosystems are interdependent across species, time, and location. Policy analysis informed by the life sciences takes account of both historical and future-oriented perspectives. Rational choice models overestimate knowledge and overemphasize economic utility as the core value, while underestimating or ignoring many qualitative human values that can heavily influence decisions, including social bonds. Choice models based on the hypothetical rational economic actor misrepresent the basis of human motivation and social ties, which skews their conclusions.

We have argued here that social science should reexamine and realign its starting point to reflect that humans fundamentally are oriented toward, and influenced by, social networks. Entanglements and emotions are factors in human choice, individually and collectively. Social animals consciously act in nonselfish ways for the perceived good of the whole, based on social and ethical considerations aside from economic, personal, or even kinship interests. Social policy analysis, therefore, should take into account how humans as part of social networks assess situations, identify core problems, and weigh alternatives. Culturally influenced definitions always help shape the perceived choices available.

Human sociability has other implications. Assumptions about human nature drive expectations about the likelihood of conflict as compared to cooperation in political life. If political and social decisions are fundamentally rooted in selfish desires, then, indeed, conflict seems an inevitable and frequent outcome. If, on the other hand, cooperation is a key element of human evolutionary success in accord with our evolved nature, then analytic approaches based on individualist and egoist premises are misdirected. The question is whether conflict is inescapable and dominant in humans and an inevitably stronger drive than cooperative urges. Primatologist Frans de Waal has noted that debates about society and government are usually built on assumptions about human nature that are not based on biology. He asserted that scientific research shows that "we are group animals: highly cooperative, sensitive to injustice, sometimes warmongering, but mostly peace loving" (de Waal, 2009). The motor of human evolution as well as human culture is more accurately identified with synergy and cooperation, not conflict. If humans naturally find ways to pursue mutual benefit, and if pursuit of cooperation and building bonds of mutual trust over time is the key to evolutionary success (see Peter Corning's argument in *Holistic Darwinism*, 2005), then seeking peaceful coexistence and cooperative arrangements sounds not utopian or foolish but practical.

Focusing on economic impacts, and on what can be monetized for the sake of comparisons (cost–benefit analysis), sets up a straw man. In order to be fully counted in the equation of costs and benefits requires the conversion of qualitative impacts (loss of species, picturesque places, clean recreation areas, etc.) to a monetized equivalent. However, qualitative human experience cannot be monetized adequately, and economic methods, therefore, result in another distortion. Translating qualitative values into monetary terms includes tools such as willingness-to-pay surveys, which are entirely subjective, ignore impacts on future generations, and do not consider cumulative impacts from multiple discrete choices. Human societies should understand trade-offs in more complex terms than estimated (monetized) costs and benefits of individual policy decisions. Social policy must look beyond the immediate issue, and consider how issues are connected to larger systems and outcomes.

Rational utilitarian analyses are also deficient to the extent that they do not calculate full social costs, particularly evaluation of long-term impacts on societal cohesion, environmental health, and sustainability. This does not mean that assessing costs and benefits is wrong-headed. It can be a useful component for decision making. Nonetheless, it should not be treated as a comprehensive or sufficient summary of the issues, or used as *the* mechanism for determining policy.

Rational choice approaches contrast markedly with the attitudes, beliefs, and political dynamics that bind people together and are undermined by a proliferation of genetic and neuroscience findings as discussed in Chapter 5. This disconnect is evident in political interactions with non-Western people. For example, seeking to build Western-style partnerships is a highly questionable strategy in areas that are dominated by kinship ties and dominance hierarchies enforced with ostracism or violence. Similarly, nation-building campaigns that assume groups will quickly shed their societal patterns and adopt new "rational" (e.g., Western) premises and loyalties alien to their culture is not in keeping with the findings of group identification and cultural studies. Assuming economic rationalism as the model for decision making in a society underestimates the power of group loyalties, shared beliefs, and long-standing bonds and animosities in framing and determining political choices (see discussion in Chapter 4 on ethnic conflict). Lessons from anthropology and ethology can provide insights into human group dynamics, including issues related to dominance and hierarchies and the importance of kinship and cultural ties, not only as mechanisms for human survival, but also for socialization and societal well-being.

Biopolicy also contrasts with determinist models, both those based on the assumption that culture/environment is the principal determinant of individual motivations and societal norms and those based on biological determinism. Deterministic approaches are not only inadequate for understanding politics and society (Thorson, 1982), but also they have all proven disastrous when applied as a formula for social transformation. Therefore, social science must supersede the false dichotomy of the nature–nurture controversy by taking into account evidence that humans are neither genetically determined nor shaped entirely by their external environment.

Environmental and genetic determinism are skewed by an overly simplistic analysis of the forces at play. While considerable research has shown the importance that the cultural and historical environments can have on individual and group choices, research and experience also make clear that biological factors, human and nonhuman, affect social conditions and outcomes. Social science must take a more balanced approach in assessing both the biological and cultural aspects of human choice and action in response to a changing environment. At their base, societal issues reflect complicated interactions between and among living systems and the environment. Sorting out important influences is further complicated by the element of human thought and technology. Determinist explanations deny that humans have the capacity to independently assess and judge situations and change direction. Humans are influenced and limited by biological as well as environmental–cultural factors, but human rational capacity has repeatedly demonstrated cultural flexibility and adaptation, the ability to supersede and transform biological and environmental conditions.

Some researchers, for example, Richard Dawkins (2006), embrace the importance of biology for understanding societal life, but simply substitute one kind of deterministic approach with another. They hold that genetically inherited traits and random genetic mutations determine human life, individually and at a societal level. Replacing mechanistic with biological determinism or reductionism does not yield an essentially better analytic alternative. A focus on human evolutionary origins can lead to an overemphasis on origins to the neglect of understanding the ends which political action should pursue. Successful reproduction, which we share with all other life forms, is neither a sufficient answer to the question of specifically human ends, nor can it support a viable ethical framework, important for establishing both justification and priority for policy decisions.

An approach that incorporates life sciences knowledge can affirm that human experience and choice affect evolutionary outcomes while at the same time shedding light on human and systemic limitations. Assessing societal issues in the context of human evolution incorporates lessons from other societies and from the ruins of past societies. It encourages recognition of the need to address social and political challenges in a way that safeguards the future and should lead to more realistic planning horizons for solving complex problems. Policy decisions benefit from a long-term perspective, in that measurable changes to natural and societal systems can require long periods of time.

Employing models from natural systems will not magically resolve our policy dilemmas, but it will provide a more realistic and fuller representation of the factors to be considered. The nature of human policy/political choice is not, and should no longer be depicted, merely as a simple matter of calculating utility or identifying inevitable forces at play.

SOCIAL POLICY AND ETHICS

Another factor missing from, or at least marginal to, social science models is the role ethics plays in policy choice. Determinist assumptions undermine the role of ethical choice while rational choice models assume that utility is the core value to be pursued. Increasingly serious problems faced by Western societies highlight the importance of better understanding not only of political processes and trade-offs associated with various policy paths but also of the role of ethics in addressing those problems.

A frequent, if not inevitable, aspect of behavioral political science is its inability to build a persuasive case for an ethical foundation for social and political life. Behavioral explanations focus more on forces affecting what is done than on human choice in pursuit of a perceived good. A policy paradigm that returns to human nature in order to understand core political goals and action, and to evaluate the likely success of competing policies, can also provide a foundation for an ethical framework. The facts we learn from nature are important in forming moral principles. For some analysts, no explanatory or research framework would be sufficient if it lacked a basis for examining and evaluating human purpose (Arnhart, 1998, 2005; Masters, 1989). Studying human evolution provides insights into factors strongly affecting what humans desire, how they seek to meet needs, and patterns of societal interaction.

Social objectives and the range of feasible actions should be analyzed in light of what is common to humans as political animals, not only of what people need to survive, but also of what supports the pursuit of excellence in human existence. Humans fulfill their needs and potential by social interaction, which occurs in a cultural and political context, whose possibilities and limits have unfolded through an evolutionary process. However, attention to human origins alone cannot fully explain or define human purpose. Human purpose goes beyond survival instincts or evolutionary selection dynamics; human society's understanding of its goals is derived from human experience over time. It arises from human biology interacting with external conditions and is manifested and developed through social experience in a complex and changing environment. Policy studies, therefore, should be grounded in a science-based perspective regarding the persistent needs, drives, and aspirations of people, individually and in groups.

Policy analysis grounded in a definition of human nature with an appreciation of biological and cultural sources of human motivation can account for the essential role of ethics in society. Society benefits by having members who can trust and support each other in seeking shared goals. The purpose of politics is to identify those shared societal objectives, to identify the common good, and to find the best means to seek it. Ethical principles are important both for describing a component of human social action and as a way of judging political actions, because choices incorporate, and are justified by, reference to what promotes the common good. Biopolicy seeks to clarify biological, psychological, social, and environmental influences on competing definitions of the social good that lead in turn to differing policy objectives and choices. Polling citizen desires at any given moment is not a reliable measure, and economic indicators likewise are incomplete in that achieving material security is not a sufficient measure of human well-being (Maslow, 1943). A healthy and stable society must distinguish between short-term desires and goals that will promote well-being over time, including future generations (this will be demonstrated in Chapter 7).

Ethics probes fundamental, theoretical, and practical questions in political life and provides a framework for addressing critical questions, such as what distinguishes human life from other life forms or the desirability of genetic manipulations affecting ecosystems upon which we depend. Policy analysis must respond to the significant and increasing numbers of important issues arising from advances in the life sciences which require ethical choices. As Roger Masters observed, "discoveries in the life sciences will force us to make decisions with awesome ethical implications" (Masters, 1989, p. 238). Social scientists should embrace the study of science issues as an important part of policy analysis and not leave these determinations to scientists, doctors, engineers, or business interests (Blank, 1982).

Social policy that is heavily dependent on rational economic man and cost–benefit analyses has numerous ethical shortcomings in addition to the methodological shortcomings discussed above. Focusing on a hypothetical rational individual for policy analysis arguably promotes not only a false but also a damaging embrace of selfish individualism and materialist desires as the basis to drive policy choices. To achieve well-being and success, social animals act with an embedded awareness of their sociocultural ties, including ethical considerations. Societal health in a democratic society depends upon making political and social decisions with reference to a shared vision of the society's needs and goals.

Using economic metrics to weigh the value of various alternatives is very difficult in complex contexts. Issues are even more difficult to monetize where risks or damaging effects of an option cannot be proven or even forecast. Estimates of future damage in natural systems (e.g., contamination of an underground aquifer or loss of a habitat for some species) are often disputed and extremely difficult to demonstrate. Giving policy weight to such estimated damages is, thus, subject to challenge. Unfortunately, often by the time harmful effects can be demonstrated, they might be irreversible. Even when damage becomes evident, proving a clear causal linkage to some event or specific factor can be nearly impossible in complex environments. This raises the question of burden of proof assumptions: should public policy regarding environmental impacts presume that action is acceptable unless harm can be demonstrated, or on the contrary, should a proposed action require some demonstration of environmental safety?

Economic analysis methods emphasize seemingly clear economic costs and gains at the exclusion of a qualitative component that may be of genuine value to all or parts of society, but which seem less certain. Moreover, cost–benefit analysis cannot adequately account for potentially severe consequences that are not readily translated into economic terms, such as loss of a pristine lake or destruction of a way of life or aesthetic treasure. Another problematic aspect is that a monetary value is put on human lives saved/lost as a way to include them in the calculation. Although including medical costs for illness and lost income is feasible, pain and suffering or family/societal disruption are difficult to measure and not economically calculable, and so might be left out of cost considerations.

Aside from inherent distortions in the cost–benefit methodologies, the highly theoretical nature of this approach does not address a basic ethical issue. Costs and benefits are assumed to be for society as a whole, and assessing whether these societal gains and losses are distributed equitably is irrelevant. If some parts of society are primary beneficiaries of the policy or activity while others bear disproportional loss or risk, the policy will still be deemed justified as representing a net societal benefit. Losses might be accounted for in the "costs" of the alternative, but that does not translate into compensation for those who actually suffer the loss. Moreover, intergenerational effects, when considered at all, tend to be heavily discounted because future costs are deemed less valuable than present costs. Relying on cost–benefit analysis also fosters a case-by-case approach that does not consider the cumulative impacts of multiple separate decisions. The logical outcome of this methodology is greater loss to be borne by future generations in order to meet current needs or desires. However, a more inclusive approach argues that human societies have a responsibility beyond the living members at a given moment in time that extends into the future and requires consideration of the well-being of coming generations (Jonas, 1984).

The ethical question regarding future generations is illustrated in the use of nonrenewable resources: depleting the resource base diminishes the options available in the future for seeking a good life, or perhaps even for basic survival (Costanza, 2014). The policy equation for using natural resources should include not only the benefits and costs (including social and aesthetic as well as economic impacts) at the time of action but also the cumulative effects on future generations. Environmental contamination has effects in the present that often continue or escalate in the future. Resource depletion typically has greater impacts as time passes; unlike toxic clean-up, the damage cannot be undone. Trade-offs between current and future use are not a new issue. An interactive approach informed by the life sciences, however, gives greater attention to the long-term implications of natural resource depletion or contaminating living systems and to associated harm to future generations. Research should evaluate who is bearing the costs, including whether they are being imposed on future generations, as well as the scope and severity of these costs, and the possibility of reversing the adverse effects.

Science research can demonstrate risks, but it cannot tell us how society should respond to the social consequences. Should society interfere when individuals choose to act in ways that increase their risks or when it creates social costs? The ethical and legal question is at what point social coercion is justified and, conversely, when interference represents too serious a violation of individual freedom. For example, some companies have instituted policies to ban employee smoking not only at work, but in general, because of proven health risks that can equate to illness and lost work days and impose higher health insurance costs on the company or coworkers. Knowledge gained from science can alter how society perceives individual free will and choice, and, in turn, can have significant effects on social attitudes and public policy.

Moreover, as noted in Chapter 1, there are many issues that are specifically linked to advances in biology where this ethical dimension is paramount. Table 1 illustrates the range of these biopolicy-based issues at the individual, societal, and global levels. A good example of the complexity of these issues and the need for a biopolitical perspective is the mounting obesity problem, particularly in Western nations.

The Case of Obesity

In recent years, the U.S. government has devoted substantial attention to skyrocketing health care costs. One of the major causes of increased serious illness (diabetes and heart disease, among others) and related health and social costs is obesity. These costs are partially borne by others and society in general. Some argue that adverse social consequences of obesity should

Table 1. Selected Areas of Biopolicy.

Individual-level issues
Human genetic intervention: Genetic counseling, carrier screening, genetic therapy
Prenatal issues: Abortion, prenatal diagnosis, reproductive technologies, fetal research, sterilization, sex predetermination
Issues within the life cycle: Nutrition and behavior, organ transplantation, psychosurgery and brain stimulation, drug therapy and usage, human experimentation, environmental mutations
End-of-life issues: Aging process, irreversible coma, terminal patients, extraordinary care, assisted suicide, definitions of death
Societal-oriented issues
Genetic diversity and human equality, population control, sex differences, race differences, aging populations, sedentary lifestyle, crowding, genetic determinants of behavior, biohazards, allocation of medical resources
Global-oriented issues
Environment, contamination of biosphere, climate change and manipulation, toxic biological/chemical substances, biological terrorism, radiation pollution, conflict/global stress, biological diversity, pollution of oceans

Source: Adapted from Blank (1982).

be borne by individuals posing greater risks and higher costs. Policies put forward to address the "obesity epidemic" all make implicit or explicit assumptions about why people are significantly overweight. The presumption often is that people choose to eat too much or eat the wrong things out of self-indulgence or poor habits or ignorance of nutrition and body chemistry and choose a sedentary lifestyle. Some of these assumed motivations have been examined in terms of market issues such as the availability and affordability of nutritious foods in some neighborhoods, or the formation of poor eating habits early in life by what and how children eat, both at home and at school.

Two very different policy responses to obesity have been widely covered by the news in recent years: one focuses on individual choice and the other on changing the social context. For instance, when several airlines announced they would charge more or require purchase of two seats for individuals who could not fit into a single seat, there was considerable backlash with some charging that such policies amounted to unfair discrimination of obese passengers. People should not be further penalized for having a biological or genetic condition causing obesity. Meanwhile, others defended the increased passenger cost, judging it fair based on airplane capacity limits for both total weight and seating space. The principle of social concern and unjust burden was pitted against the equity argument that those needing extra service (space in this case) should pay for it, rather than the service provider or society as a whole. Others voiced concern that causal factors are complex and society becomes too oppressive and potentially unjust in placing responsibility on individuals for their conditions.

Another example of policies aimed at reducing obesity was the move by then New York City Mayor Michael Bloomberg to forbid the sale of supersized soft drinks in the name of public health. The rule was immediately attacked as a "nanny-state" intervention and a violation of individuals to make their own choice. Beverage makers and business groups challenged it in court arguing that the mayoral-appointed health board violated the state principle of separation of powers. It was rejected by a unanimous state appeals court, which said the mayor had overstepped his authority in trying to impose the ban to combat obesity.

Critics note that if society wants individuals to pay the cost of their excesses – for example, higher insurance costs or paying for extra space on a plane – perhaps it should distribute part of the responsibility to producers of unhealthy foods and additives, or to advertisers who persuade people to adopt unhealthy habits. In these two initiatives, opposition was leveled often in the name of respect for individual freedom of choice. Arguments in favor of unrestricted choice, however, should not only be weighed against public health needs and associated societal costs of obesity, but also in light of how manufacturers formulate and market such drinks to the public: what individuals "choose freely" to consume must be tempered with knowledge of how large-scale manufacturers manipulate markets and target psychological appeal to increase product adoption. Furthermore, the addictive nature of sweeteners should also be considered (Freeman, Cummins, Tucker, Ostland, & Hock, 2011).

These approaches represent limited attempts to address social costs from obesity. Private sector policies tend to try compensation to readjust the costs generated by the situation. Public policy should meet a different standard of fairness, taking into consideration the root cause of the situation and try to mitigate or bring relief. Solving a problem is preferable to mitigating the damages from it; preventing illness not only reduces human suffering but also has lower social costs than compensating for the health problems. Crafting effective solutions is an important criterion for good public policy, and such solutions require input from the life sciences.

There is a direct connection between body chemistry and nutrition, and mental as well as physical health. Recent scientific research indicates that there are other factors that might affect evaluations of what justice or equity require and what kinds of corrective action are more likely to be effective. Studies have found important biochemical differences related to obesity; some individuals lack a biological signal of satiation that clues the body to stop eating. A recent study suggests that overeating, like drug addiction, can dull the brain's response (the dopamine reaction), and it takes more food, more fat to satisfy hunger. Seeking an efficacious policy would, in this case, call for a therapy to rebalance the dopamine response (Gelling, 2013). This research illustrates that important social issues with significant political consequences have very direct ties to biological or biochemical systems, and that addressing them effectively might require medical interventions rather than selective taxes, higher prices, public pressure, or education to bring about change. Justice and efficacy are linked in that a just solution should in fact address the problem at hand.

INTEGRATION OF COMPLEX INFORMATION SOURCES

One of the many scholarly and policy contributions of Lynton Caldwell was his early appreciation of the importance of the life sciences to political science and policy (Caldwell, 1964). He recognized a mounting need to conduct multidisciplinary research in order to better understand issues and conditions affecting human society and to identify those factors that need to be managed or contained and those that might be changed. Policy analysis needs to integrate complex sets of information and, if it is to be practical, must distinguish issues that can be readily solved from those that require long-term management or those that demand adjustments to human activity. As societies struggle to define and establish sustainable practices, they have a pressing need for reliable scientific data. A successful strategy requires that society balance current needs and/or desires and profits against future costs or shortages. Policy makers must separate genuine needs from preferences and convenience. They must also determine how much potential damage is acceptable and when to curtail or eliminate production, products, or activities.

The life sciences have brought about fundamental changes in the way we understand some social phenomena, which, in turn, can alter how we define viable policy alternatives. Sound policy and law must be built on realistic expectations about how or why people act, which is also important to successful leadership, community-building, and strategies for dealing with broad societal problems such as crime and violence. As discussed in detail in Chapter 5, substantial research on the functioning of the human brain provides important hypotheses and information about human decision making. In addition, brain mechanisms and conditions are highly linked to self-control and with antisocial and aggressive behavior (Blank, 2013; Ellis, 2003). For instance, if correlations between exposure to certain toxins (e.g., lead, a known neurotoxin) and higher rates of criminal and violent activities (Masters, 2002, 2003) are confirmed not only as a strong association but as a causal link, it would have direct public policy implications. Programs aimed at reducing crime would need to focus more attention on reducing or mitigating certain chemical exposures. Similarly, the biosciences can provide crucial knowledge about addictive behaviors, including perhaps compulsive eating, which are related to antisocial as well as self-destructive actions.

Research findings related to decision making, coalition-building, and leadership benefit from a biobehavioral perspective, including an examination of parallels between primate behavior and human group dynamics. Evidence from studies of primate cooperation, submission, and dominance provides valuable insights for human group dynamics, group loyalties, as well as exoteric and esoteric aspects of leadership. Greater attention to the effects of culture and kinship on decision making and legitimacy in traditional societies can shed light on the implications of weakening human bonds, individual and societal, that can threaten stability and legitimacy in societies that contain large numbers of disconnected people.

The Case of Public Health

The benefits of bringing life sciences knowledge to public policy discourse is very clear in the field of public health, where research has had substantial impacts on social policy. Policy decisions regarding science funding reflect and also help define the kinds of research that is deemed most likely to lead to new capabilities, cures, or treatment for serious illnesses such as diabetes or cancer. In 2013, the U.S. government announced an expanded 2014 Brain Initiative research program that has physical, biological, social, and behavioral components. The major objectives of the initiative are to increase knowledge and understanding of how the brain functions (storage, processing, and decision making) and discover mechanisms related to brain disorders and injuries that participants expect will yield policy benefits (The White House, 2013).

As knowledge of how diseases are transmitted and affect individuals and populations has heightened, these findings often result in new societal responses regarding the care of individuals and communities. In Western societies, attitudes toward and management of mental illnesses have changed dramatically based on improvements in neuroscience as well as psychology, biochemistry, and pharmacology. Western societies no longer shun people with epilepsy as possessed by evil spirits and rarely quarantine individuals or communities out of the fear of spreading contagious diseases (Centers for Disease Control and Prevention, 2014). Even the widespread fear of AIDS in the 1980s, despite calls for isolating AIDS carriers, did not result in quarantine policies. One century after Typhoid Mary's 30-year isolation, the first U.S. forced quarantine in over 40 years of a contagious disease carrier triggered a national debate on a justifiable cause for forced coerced quarantining as a policy (Altman, 2007; CIDRAP, 2007). Therefore, it is clear that science research under favorable circumstances can lead to more targeted and socially effective policies.

Given the staggering and still mounting national cost for health care, especially in the United States, funding of public health policies always entails trade-offs (Lamm & Blank, 2007). Therefore, society should be fully aware of the ethical choices implicit or explicit in the medical practices it supports and who is eligible. Since the health care system can never meet all health needs, much less social needs that have a health component like obesity, policy priorities, by necessity, must be based on some allocation criteria, and those criteria should be based on scientific findings and ethical grounds as well as affordability or other societal concerns.

Life sciences data can improve policy processes and decisions. Policy analysts should look to both processes and outcomes in setting agendas, identifying goals and options, conducting alternatives analyses, and evaluating consequences. For example, if we consider the process for setting standards for drinking water contaminants and the addition of fluoride to drinking water, one can evaluate whether the process by which it was approved followed typical procedural requirements, or scrutinize the quality of data/scientific evidence used, the extent to which risks were considered, and how any negative outcomes were considered and addressed (Coplan & Masters, 2000; Masters, 2003). Similarly, setting priorities in agricultural policy, which greatly affect U.S. farming practices, should also be concerned about greenhouse gas emissions. University of Michigan researchers found that, in differing global locations, the frequent practice of adding excess nitrogen to the soil as fertilizer causes quickly increasing formation of nitrous oxides, the third most potent greenhouse gas (after methane and CO_2). Nitrous oxides also destroy stratospheric ozone (Shcherbak, Millar, & Robertson, 2014). Biopolicy does not provide the answers but a holistic perspective helps ensure that issues are properly framed, bringing scientific findings as well as political realities to the analysis.

These same principles, of course, are important for natural resources planning, transportation policy, and trade policy. Current and projected population needs for critical resources such as clean water should be compared to data on national and regional carrying capacity to gauge whether supplies are likely to be sufficient, given current projections of usage, the effects of climate change, and increased occurrence of violent storms. Policy analysts and politicians recognize that they need basic information to address natural resource needs as well as agricultural, medical, and health issues in domestic and foreign policies. Nevertheless, neither the political system nor academe has fully embraced the importance of more completely integrating life sciences information into the study of politics as well as the policy process. Before concluding this chapter, we examine in more detail the contentious and complex issue of hydraulic fracturing.

BIOPOLITICAL PERSPECTIVE: ENERGY POLICY AND HYDRAULIC FRACTURING

As noted earlier, protecting long-term environmental health could be substantially improved if issues were understood and presented in the context of ecosystem effects (see Chapter 7 for more details). Increased knowledge about human and ecological evolution might encourage policy analysts and policy makers to adopt a long-term ecosystem perspective when addressing energy policy.

Because knowledge from the natural sciences has influenced U.S. environmental laws and practices, there have been dramatic improvements in safeguarding the quality of air, water, and land. Energy policy, however, continues to generate heated legal and policy battles over the life cycle risks and damages associated with various forms of energy development, and whether to require new or stricter industrial standards. These decisions must rely on increasingly refined and complex levels of analysis. The goal for policy makers and courts is to safeguard public health and the environment, while at the same time protecting the viability of industries and freedom of the marketplace from unnecessary restrictions.

Hydraulic fracturing (fracking) for natural gas deposits locked in shale formations illustrates the intersection of life sciences information and public policy, and the tension between protective policies and pressures to minimize restraints on industrial production. Fracking also highlights some of the limitations of cost–benefit analysis as a primary tool for decisions and the ethical choices that underlie policy.

Fracking relies on highly complex technologies developed over the past half-century, although some have been widely used for only about a decade. Proponents point to a large number of benefits that accrue to development of gas resources made possible by fracking. Those opposed, or with more cautious perspectives, point to evidence of contamination from shale gas development and emphasize that potential risks to water, air, and land, as well as myriad social costs, are serious. Contamination of groundwater, in particular, could have catastrophic and irreversible effects. Therefore, many interests want to slow or stop the rapid development underway, at least until more is known about environmental, health, and societal effects.

Benefits of Fracking

Development of domestic natural gas resources promises many benefits, providing both domestic and international political advantages related to increased energy independence, job growth, and a cheap domestic energy supply. U.S. Steel Chairman John Surma likened fracking to the economic engine of coal production nearly 200 years ago (Mufson). There are strong economic incentives. Domestically, it promises to provide a cheap source of energy for decades and increase tax revenues at all levels of government. Supporters of aggressive development claim that it creates millions of U.S. jobs, boosting or even reviving local economies in the areas with gas resources, as well as giving much needed business to feeder industries across the nation. Lower energy prices for domestic industries have increased their international competitiveness and convinced some offshore production to return to the United States.

Expected benefits, however, are based on assessments that are not universally accepted. A key component of the benefits equation is the size of the untapped gas sources and what proportion is recoverable. Estimation methods are a complex matter and results are, like most issues related to fracking, contested. Nation-wide estimates of a 90- to 100-year gas supply, for example, have been frequently repeated and are supported by Department of Energy (DOE) estimates, but rejected by other experts as being far too optimistic. In 2012, the DOE revised government estimates downward (EIA, 2012, pp. 56–60), although they still significantly exceed projections by some experts (Fuquay, 2013; GAO, 2012, pp. 16–25; Hughes, 2012). Gauging the overall supply is also important for decisions to build the very expensive infrastructure for liquefied natural gas (LNG) export, which requires long-term cost recovery to be profitable, as well as government approval of new pipelines (requiring right-of-ways) to LNG platforms. Aside from added pollution and risk that could result from exporting gas, the benefit claimed for cheap energy is undercut if export raises domestic prices, which it is meant to do.

Fracking results in the production of cleaner burning natural gas fuel, with 50–70 percent reduction in greenhouse gas emissions per unit of energy as compared to coal. However, claimed environmental benefits are challenged. One challenge focuses on net benefits for climate change, based on a life cycle assessment of pollution, including methane leaks ("fugitive" emissions), throughout the extraction/processing/transport and use of natural gas. Since methane is a very potent greenhouse gas, far more potent than CO_2, significant fugitive emissions can outweigh the benefits of gas as a cleaner burning fuel during actual use. No comprehensive data exist on the amount of methane escaping at the tens of thousands of well sites, mostly during extraction, or during processing and transport. Experts differ widely about which sources and data are most representative and which models to use in estimating national emissions. As a result, some life cycle assessments of natural gas energy show dramatically less environmental benefit or even that gas is not preferable to coal in terms of total greenhouse gas effects (Alvarez et al., 2012; EPA, 2013; Howarth, Santoro, & Ingraffea, 2011; Hughes, 2011; O'Sullivan & Paltsev, 2012; Tollefson, 2013).

Another challenge questions the expected impacts of natural gas production on overall energy production and use. While some see cleaner gas as a bridge fuel to replace coal, others contend that decades of cheap gas could cripple newly emerging renewable energy companies and associated research during a critical phase of development. Another concern is that climate change benefits will not occur, because coal not used in the United States is likely to be sold abroad, with little or no net reduction of greenhouse gasses globally.

From a foreign policy perspective, natural gas production has several important benefits. Significantly improving U.S. energy independence could reduce the need for its involvement in volatile Middle East politics and protection of international oil routes. It would make the U.S. economy less dependent upon fluctuations in international oil markets and improve our balance of payments. It is also desirable as a means of providing European allies an alternative to reliance on Russian gas, if Russia persists in using gas resources as a political tool. Russia's bold interference in Ukraine has increased the attractiveness of this argument. Reducing U.S. greenhouse gas emissions would also increase U.S. credibility on global warming issues and ability to attract international support for proposed solutions. Industry LNG advocates note that exporting cleaner burning gas fuel will promote the Administration's climate agenda in addition to creating jobs, improved balance of payments, and assistance to European allies.

Risks of Fracking

Multiple concerns have been raised about the short- and long-term impacts this industry might have on health and the environment (e.g., summarized in GAO, 2012, pp. 32–55; EPA IG Report, 2013). The industry and many federal and state officials contend gas development is safe when done properly, and although faulty wells, company errors, dumping, and accidents have caused some problems, the industry has a reasonably good record thus far. Some critics, however, give more weight to contamination incidents and have concerns that oversight is insufficient to identify and control problems. They also cite the lack of data and the difficulty of proving a source of contamination without baseline data. More importantly, opponents emphasize inherent risks, including significant releases of toxins to air, land, and water, as well as inevitable human and materials failure rates along a chain of complex industrial activity, which is likely to result in long-term, irreversible contamination of water and land, causing regional or ecosystem health effects. For these concerns, a lack of clear data proving current damage does not provide evidence of safe processes and practices.

Consensus on risks will not be possible without an agreed upon health impacts assessment, which is lacking primarily because data are sparse, but also because of debates about which data, geological conditions, or industry practices might be used to estimate exposures from gas development in a region or nationally. Geology and industrial practices can differ significantly from well site to well site. As noted above, the likelihood of broad ecosystem impacts through contamination of surface and groundwater, as well as from air emissions and inadequate waste management, is vigorously debated. Establishing consensus on how to use existing, or design new, data collections to assess overall health impacts is unlikely in the near future. One human health effects study noted the breadth and complexity of the issues, arguing that health effects go beyond exposure to toxins, that "oil and gas boom and bust cycles have deleterious effects on the psychosocial welfare of a local population," which should be taken into account (Witter et al., 2008, p. 39).

In addition to health and environmental impacts, communities are worried about social disruption and discord, loss of local businesses dependent on environmental health, burdens on local infrastructure, and aesthetic degradation. Potential risks are of greater concern if industrial development would threaten ecologically sensitive areas, such as the watershed providing drinking water to millions of people in New York City, which currently does not have a treatment system for drinking water.

From extraction to transmission and disposal of wastes, gas production can pose risks to air, land, and water. In the *extraction process*, protecting water resources, particularly drinking water, has been the central focus for environmental concerns. Fracking requires very large amounts of water; estimates vary, but generally fall in the range of two to eight million gallons of water for each fracking process (e.g., USGS, FAQs, 2013), with multiple fracks per well. In 2012, tens of thousands of wells were being drilled each year in the United States (EIA, 2012, p. 59). Water used for fracking (frack water) is laced with toxic chemicals and generally cannot be used for other purposes, and recycling frack water for use at other wells is limited. Water usage is of particular concern in dry or drought-stricken areas.

A persistent issue is whether frack water, with additives including highly toxic chemicals, will seep into aquifers and contaminate them. Frack water has contaminated ground and surface water because of faulty well construction, overflows, leaks, and accidents. Contaminated surface water poses risks to plants and animals as well as drinking water sources. Toxins released at well sites can travel through the environment, resulting in broader ecosystem contamination, with both long- and short-term effects.

Air emissions during *extraction and transmission* of gas include not only methane but also volatile organic compounds, which pose risks to human health that could be significant near well sites (Gilman, Lerner, Kuster, & de Gouw, 2013; McKenzie, Witter, Newman, & Adgate, 2012; Petron et al., 2011; Witter et al., 2008). There are calls for additional studies and monitoring of those emissions. Air pollution associated with the huge volume of truck traffic and with building and managing the well site (which includes a variety of tanks) are other sources of exposure.

Waste transport and management is needed for liquid and solid wastes. Large amounts of water flow to the surface after fracking operations, both returning frack water injected for the extraction process, and groundwater containing toxic chemicals and minerals, including dense brine, often with radioactive elements. The contaminated waters pose risks related to improper handling onsite (unlined or overflowing holding ponds), spills, leaks, or accidents in transport and improper or inadequate disposal.

Available disposal methods for wastewater all have associated risks, and there are no comprehensive disposal standards for it (Hammer & VanBriesen, 2012). Deep well injection, the most frequent method, has a demonstrated relationship to seismic activity; at least a few earthquakes have caused surface shaking and damage. Wastewater has been trucked to treatment facilities not equipped to remove all of the toxins in this waste, which can damage treatment processes in addition to releasing contaminated waters into the environment. Open air evaporation, burying in shallow pits, spraying untreated wastes on roadways as a de-icing agent, and some instances of illegal dumping have also been documented.

Solid wastes from drill cuttings and muds are sometimes buried onsite, but are often sent to landfills; whether contaminants leaching from these wastes pose a threat has not been assessed. Levels of radioactivity in wastewater and solid wastes, and possible exposures, are not well understood. Studies from different shale formations indicate that levels of radioactivity in well materials are far higher than previously believed (e.g., Climate

Connections, 2012; Rich & Crosby, 2013). Large volumes of waste are transported often to considerable distances, which increases traffic, noise, and potential accidents and spills.

The overall tally of local risks and benefits makes clear only that there are many unresolved issues. Some have benefited from an economic boost from gas production. Others have complained of contaminated drinking water, significant health impacts, erosion, or environmental degradation of land or water resources that threatens their livelihood, including livestock illness or death. Since there is little baseline data from the areas where fracking has occurred, even near-term analytic results are debated. Many factors contribute to environmental conditions, so it is not surprising that studies have differing results. Industry practices and state oversight vary and external conditions related to geography and weather can minimize or magnify pollution. Longer term consequences to groundwater are a significant concern, for which no data exist to resolve differing assessments.

Aside from health and environmental risks, fracking for shale gas is a large, dirty, noisy, and sometimes deadly industrial process. It dramatically affects the quality of life in the community. Impacts on local infrastructure, such as deterioration of roads due to truck traffic (which also increases air pollution significantly, affects safety, and creates noise, congestion, accidents) are common problems. Disruption and discord can arise from disagreements over allowing fracking in the community, or the divisions between those who gain financially and those who do not. Businesses depending on the guarantee of a clean environment – tourism, organic farming – are at grave risk. The influx of outside workers brings higher crime rates and stresses on local services, including health care, hospitals, police, food, lodging, and entertainment, which also raises local prices (Pressconnects.com, 2012; Stedman et al., 2012). The boom–bust cycle brings numerous long-term problems, including rifts in society, psychological effects, and excess capacity from the business surge which then diminishes or disappears.

Public policy decisions typically weigh the risks associated with an activity against the benefits expected, and such analysis provides valuable information for decision making. Not only are the risks of gas development through fracking controversial and unresolved, but projecting environmental benefits from natural gas development is also a complicated and contentious matter. There is a lack of data to address critical issues. A list of missing or contested data mirrors the long litany of risk–benefit categories, from the volume of recoverable natural gas in shale formations to potential long- and short-term ecosystem effects to infrastructure damages to human psychological effects of boom–bust cycles. More data could help resolve some issues, but because formations, well constructions, industry practices, and state government oversight vary in relevant ways from area to area, better data are unlikely to settle risk estimates for those on opposite sides of the issue.

A holistic and systemic approach might favor a slower, more measured approach. Proponents argue the need for energy development now, as if the resources will disappear if not extracted in the immediate future. Data on methane releases are central to assessing the true net benefits of burning gas over coal. Policy evaluation should address the likelihood that U.S. coal would be shipped abroad and burned, and how that affects the risk–benefit equation. The impacts of developing LNG capacity are another area needing greater clarity, regarding the environmental impacts, effects on domestic consumption, including higher prices, and any net climate benefits.

A biopolitical perspective should take into account who is bearing the risks and who is reaping the benefits. Even if overall risks are outweighed by the benefits of gas production, justice demands that society not pursue a policy whereby risks are borne by the few and/or by future generations, while a small group garners most of the benefits. A mere totaling of estimated risks and benefits, as noted earlier, fails to take into account that those who stand to reap huge monetary benefit are frequently quite different from those bearing the losses/risks (Kelsey, Metcalf, & Salcedo, 2012). A life sciences approach should highlight if risks are disproportionately distributed, particularly if the "cost" includes ecological disaster in some areas or for future generations.

CONCLUSIONS: HUMAN-CENTERED BIOPOLICY

Advances in the life sciences are presenting us with critical choices, which will profoundly affect society. These choices should be articulated in terms of the risks, opportunities, and impacts in the near term and long term. Each requires an evaluation of options and related outcomes, including ecosystem, biological, and biochemical impacts on society and the planet. As more of these kinds of issues face us, we do not seem to be making much analytic or political progress in addressing them. Premises and beliefs about politics inform not only the range of policy options defined but also how laws are interpreted and how legal precedents apply to new circumstances. As Caldwell commented years ago, we are not prepared to cope with the problems; our political culture is very reluctant to face the true issues (1972, pp. 44–45). It is still the case that few in the government seem to be familiar with the scientific factors underlying the complex of biotechnical issues facing society (Funke, 1985). At least part of the problem, however, is the failure of social science to speak clearly and persuasively about the issues and what is at stake. Continued reliance on mechanistic, deterministic models threatens to make political science largely irrelevant, in that it will not offer vital insights and policy choices or contribute to the important policy debates. Interdisciplinary learning would help build a new paradigm for policy. These issues are not just matters for scientists; they are critical policy matters that need greater attention from policy analysts and greater accountability from manufacturers, public officials, and politicians.

Policy analysis is "a critical linkage not only with other subfields in political science but also with other disciplines" (Blank, 1982, p. 40). Political scientists should be well positioned to address these complex and important issues. We need more sophisticated analyses of the role of technology in addressing, and sometimes creating, problems. Political scientists should understand technologies, their costs (especially public costs), associated risks, consequences of broad use, and trade-offs. How does society draw the lines between what government prohibits compared to what it ignores or tolerates and what it supports, directly (e.g., farm subsidies, research funding) or indirectly (tax breaks, regulations, education, etc.)?

Future research should focus on and anticipate emerging issues and pending crises, calling for massive global adjustments to political disruptions arising from religious and cultural divides, as well as to natural disruptions from climate change and to technologies with novel and far-reaching capabilities. Domestically and internationally, nations face multiple levels of issues arising from environmental and related public health challenges. To prepare for dramatic population, agricultural, public health, and ecological impacts will require substantially greater attention by social scientists to complex biological and environmental as well as political systems. Policy analysts and policy makers must also seek a better understanding of how human populations can best respond to a variety of far-reaching and overlapping changes. In the United States, climate changes are already starting to necessitate new considerations for public policy (e.g., adequate water supplies or persistent flooding) and new constraints on individual and business decisions that have heretofore been relatively uncontrolled.

Failure to distinguish opinions from verified facts can allow strongly asserted beliefs or self-serving rhetoric to justify policy, regardless of weak or false premises. It increases the risk that we will pursue policies that lack substantiating evidence, or worse still, pursue policies despite evidence to the contrary. In recent years, we have numerous examples of leaders espousing positions with regard to environmental controls or the need for fiscal controls or the likelihood of some consequence that have ignored or dismissed pertinent information, basing policy decisions on ideology or political convenience instead. Good social science analysis does not solve the problem of poor leadership, but it can provide a counterweight to false assumptions and shed light on core factors that should be addressed.

One objective of policy analysis is to understand political motivation and action, some of which is founded in biology, some of which is culturally learned, and some of which is a complex reaction to a multidimensional and changing environment. Biopolicy requires that we move away from social science analyses that do not start with the social nature of human animals that are part of and not separate from the natural world. From a biopolitical perspective, human action is subject to more conditions and limitations than many conventional conceptions of free will imply, but human will and ethical commitments are factors in assessing policy choices. In seeking to understand core political goals and action, and evaluate the likely success of competing policies, a biopolitical approach reveals the need for ethical inquiry and insight. Especially in multicultural societies, with competing ideologies, the need for a shared frame of reference to define and defend policy directions is ever more evident. Shared values and goals are an important basis for domestic as well as international cooperation. Understanding humans from an evolutionary perspective provides insights into human pursuits, not just for survival, but to achieve higher human ends that are deeply rooted in the evolution of social life; that is, humans look beyond material goals to seek happiness.

Emerging biopolitical problems highlight the importance of anticipatory policy and of assessing political processes and trade-offs associated with various policy paths. Finding workable policy solutions not only depends on human cooperation (political feasibility), but also on accurate expectations regarding the limits and possibilities of external physical systems. There are important differences in the interactions among biological life and the surrounding cultural and physical systems occurring at various levels of human society and across widely varying populations at any level. Constraints and impacts differ at each level and this requires clear methodological consideration for each level (Masters, 1989, especially chapter 4).

Studies of individuals or groups will differ from evolutionary considerations at the population or species level (Blank & Hines, 2004, especially chapter 4). Holistic analysis respects complexity in pursuing an integrative approach and assessment of relationships across levels of analysis, from individuals to communities to the international community.

The fusion of social science and the life sciences should inform basic policy considerations, including the substance of the policies as well as the processes for their identification, proposed resolution, and implementation. Building on insights from the life sciences and the evolutionary theory, multiple lines of research might be developed in a wide variety of policy areas.

It is a hopeful sign that many political scientists in the past couple of decades have been giving increased research attention to important life science issues, particularly issues surrounding public health and medicine and the environment. Whether or not they associate their work with biopolicy, they are taking a more holistic approach to policy analysis and giving critical attention to the use of life sciences research to inform public policy decisions.

Addressing these issues more thoroughly will require a broader incorporation of life sciences studies in the social sciences curriculum and a move toward cross-disciplinary studies. Policy analysts and decision makers urgently need this knowledge to understand the options and trade-offs. Citizens should have a basic literacy in life sciences concepts to judge the plausibility and desirability of policies. Lacking familiarity with life sciences concepts leaves the general public too easily manipulated by biased and ill-informed sources. Breaking with existing patterns is a very difficult enterprise, and one opposed by those with a stake and status in the existing structures. The heightened need to embrace an interdisciplinary perspective poses complications and challenges to the status quo, including the structure and rewards system embedded in university departments and disciplines. It will also require a basic change in our universities that removes barriers and penalties associated with crossing disciplinary boundaries, for students and especially for professors (for more details on this, see Chapter 8). Interdisciplinary approaches should be expanded and passed on to the next generation of scholars, public policy officials, and activists. Premises and beliefs about politics inform not only the range of policy options defined but also how laws are interpreted and how legal precedents apply to new circumstances. In short, interdisciplinary learning would help build a new paradigm for social science policy research.

CHAPTER 7

ECOLOGY AND ENVIRONMENTAL ISSUES AT THE GLOBAL LEVEL

Samuel M. Hines, Jr.

> The world in which we now live is a world of very obvious danger and potential extremity, the first epoch in human history in which it was readily open to millions of human beings to see the earth over time as a natural habitat perilous for the entire species to which they belong and, above all, the first epoch in which men and women could soberly confront the possibility of a natural end to their own species brought about through human action itself. These dangers, to be sure, for most of us bewilder more than they enlighten. The passion of fear on which Thomas Hobbes hoped to build so much is here woven too deeply into the occasions for experiencing it to lend us any very enlightening guidance on how to lower it to more tolerable levels. The classical preoccupation of political theory with legitimate rule, a domestic social good, and the casuistry of just or unjust warfare are not comprehensively irrelevant to this menace. But they fall rather obviously short of showing us how to confront it.
>
> John Dunn (1990, pp. 196–197)

> It would be wise to observe the following principle: if we save the living world, we will also automatically save the physical world, because in order to achieve the first we must also achieve the second. But if we save only the physical world, which appears our present inclination, we will ultimately lose them both.
>
> Edward O. Wilson (2012, p. 294)

Politics and the Life Sciences: The State of the Discipline
Research in Biopolitics, Volume 12, 205–228

ISSN: 2042-9940/doi:10.1108/S2042-994020140000012006

INTRODUCTION

Whether they prove to be timeless or not, most political theories originate as a response to a particular crisis or problem (Wolin, 1960). Today those crises and problems relevant to any discussion of ecopolitics and environmental politics and policy include population growth; constraints on vital resources (e.g., water, fisheries, arable land); industrial pollution; the waste from weapons of mass destruction; the potential devastating effects of biochemical weapons; biotechnology and genetic manipulations; resistant strains of bacteria; and the most controversial of all today, climate change and global warming (see Lomborg, 2004, for a more comprehensive list of global crises). E. O. Wilson, in lamenting the loss of our biodiversity, notes the rapid disappearance of tropical forests and grasslands and other habitats where most of the diversity of life exists (2012). He cites the changes caused by "HIPPO" (Habitat destruction, Invasive species, Pollution, Overpopulation, and Overharvesting, in that order of importance) and speculates that half of the species of plants and animals might become extinct by the turn of the next century (Wilson, 2012, p. 294). Likewise, Cairns (2001) reminds us that the pursuit of sustainable growth, even the so-called "smart growth," absent consideration of and action to mitigate the negative long-term effects to the natural environment, could result in catastrophe, for example, overshooting carrying capacities of the planet. These crises call into question the capacity of humanity to address these problems in a timely and effective manner. Clearly, what is called for is a holistic approach, based on a full understanding of what light contemporary science can shed on these issues and on the ability of governments and nonstate actors to address them.

We begin this discussion of environmental and population issues in global politics by emphasizing the direct relevance of the life sciences and the evolutionary theory to the macro-level issues associated with the current global environmental crisis. Our contemporary crisis, variously described, is fundamentally (though not exclusively) an ecological one and what most needs to be addressed from a life science perspective (biopolitics) at the present time are those political and economic strategies that may no longer serve as successful, long-term survival strategies for our species.

Advocates of Green politics have characterized our current crisis in the following way:

> The starting point of Green politics is the recognition that we find ourselves in a multifaceted, global crisis that touches every aspect of our lives: our health and livelihood,

> the quality of our environment and our social relationships, our economy, technology, our politics – our very survival on this planet. The nations of the world have stockpiled more than 50,000 nuclear warheads, enough to destroy the entire world several times over, and the arms race continues at undiminished speed. While world-wide military spending is more than one billion dollars a day, more than fifteen million people die of starvation annually-thirty-two every minute, most of them children. Developing countries spend more than three times as much on armaments as on health care. Thirty-five per cent of humanity lacks safe drinking water, while nearly half of its scientists and engineers are engaged in the technology of making weapons. Economists are obsessed with building economies based on unlimited growth, while our finite resources are rapidly dwindling; industrial corporations dump toxic wastes somewhere else, rather than neutralizing them, without caring that in an ecosystem there is no "somewhere else." Modern medicine often endangers our health, and the Defence Department itself has become a threat to our national security. (Spretnak & Capra, 1985, p. xv cited in Carter, 1999, p. viii)

Kaplan (1996, 2012) describes the devastating conditions of poverty and disease in Africa and other parts of the world. Moreover, Hertsgaard's Earth Odyssey (1999) portrays in stark detail the depressing realities of environmental degradation, illness, and poverty. Moreover, he underscores the unevenness of this global problem (1999, p. vii). If poverty (and its relationship to the causes of inequality) is the biggest environmental challenge of our time, the unbridled pursuit of wealth and its concomitant acquisitiveness for material consumer goods has become the biggest ecological/ environmental burden. The consumption patterns of the nearly one billion people who live in the affluent world of Europe, North America, and other industrialized countries cause much more environmental damage – more greenhouse gas emissions, more forest cutting, and more soil, air, and water pollution – than do the strivings of the impoverished human majority (for the underlying population dynamics, see Cohen, 1995).

China clearly illustrates this point. Measured by population, Chinese outnumber Americans nearly five to one. Until now, however, the United States has dwarfed China's total environmental impact because Americans consume 53 times as many goods and services *per capita*. That is not to say that China does not have serious environmental problems, especially frequent debilitating pollution, but also potential carrying capacity issues that could significantly impact its continued economic growth. The rapid economic growth in China over the past two decades as they have participated extensively in the global market economy has brought new challenges that Chinese leaders are only beginning to identify and address.

William Ophuls (1977, 1997) has argued that the liberal foundations of modernity as they have influenced modern politics must be found wanting in the context of our current predicament. The overreliance of liberal democratic capitalist systems on individual strategies of consumption (Leiss, 1976; Wachtel, 1998) and the resultant exploitation of the natural environment are proving to have profoundly negative consequences for the natural world and, thus, for our species as well as other species. One ramification is the loss of biodiversity highlighted by Wilson (1988). Hayward (1998), in advocating a political theory informed by ecological values, has taken the position that we must find ways to integrate ecological values expressed as interests into our already self-interest-driven policy processes.

These two theorists, Ophuls and Hayward, can be seen as examples of what Stavrakakis (1997) differentiates as ecologism and environmentalism. The former refers to a more extreme and thoroughgoing ideology (that, for some, calls for an expanded role for state control and even authoritarian politics), while the latter reflects a more incrementalist model of politics where environmental concerns are addressed, but only piecemeal over time through liberal democratic capitalist systems and in contention with other goals and values (see also Eckersley, 1992, 1998). One goal of biopolicy (Caldwell, 1987; Funke, 2013) would be to replace current approaches to public policy with a policy framework that takes the life sciences seriously and seeks to create policies that pursue the common good and address issues of fairness, as understood by Corning (2011a, 2011b), rather than policies that result from the success of special interests in pursuit of their own policy agendas.

We are less concerned here with debates about the important differences in these ideological positions (but see Ferry, 1995, for a critique of "deep ecology" and its ideological implications) and more concerned with the significance of persuading political scientists and policymakers of the need to think ecologically. This requirement means reasoning more synthetically, holistically, and systemically rather than thinking analytically based on a mechanistic worldview, especially one that considers the natural environment as subject to the control of human beings and as a limitless resource that is ours to exploit. For an attempt to develop a typology of the political theory of human's position relative to the natural world and their attitudes toward the natural world, see Masters (1989).

The British political theorist, John Dunn (1990), would have developed a modern sense of prudence that goes well beyond instrumental rationality to embrace the need for a reconceived notion of political community. We agree with Dunn and would assert that we are especially in need of that quality of vision which has often been associated with the most time-honored works on political theory (Wolin, 1960). There is an urgent need to embrace a new biopolitical theory of a global ecopolitical community that is both compelling and rooted in a contemporary understanding of evolutionary biology and ecology. In other words, we desperately need ecological prudence across the world today. As is common to the three biopolitical theories of Masters, Schubert, and Corning, a life-science-based approach to political science would address all levels of reality and be capable of studying those phenomena with various methodologies. The interactive paradigm requires that we give attention to genes, the body, groups, *and the environment* and ignore none of them and, furthermore, that we seek to understand their interactivity (see Figs. 1 and 3 of Chapter 2).

Although we cannot hope in this chapter to contribute significantly to the larger project of articulating an ecologically based policy agenda set in the context of the biopolitical theory, we delineate several lines of inquiry and areas of priority for public policy that are essential to that broad endeavor (Pirages & DeGeest, 2004). There are three distinct lines of contemporary inquiry that can be drawn upon to explicate the political issues raised by current environmental crises and the "climacteric" (see below). The first of these is the biopolitical theory, which has been presented earlier and will be the focus of this chapter. The second comprises the rational choice theory and the economic theory, joined with utilitarianism and liberal democratic capitalism (see Kassiola, 1990), that are representative of mainstream Western political ideology. Finally, the third approach is the postmodern critical theory, which generally reduces to subjectivism and is, therefore, not discussed at any length here. Nor do we analyze Marxist-socialist theories, even though contemporary work by its proponents has much to say about the defects of liberal democratic capitalism and the impact of the wealthiest countries on the poorest countries, albeit with little emphasis on environmental issues. Predictably, we argue that a biopolitical perspective informed by the evolutionary theory and ecology is preferable and offers genuine possibilities for characterizing the problems we face and challenging many of the prevailing assumptions that guide contemporary policy making.

AN ECOLOGICAL CRITIQUE OF MODERNITY

Any attempt to characterize, much less critique, modernity is fraught with extreme difficulty (see Toulmin, 1991). Nonetheless, some characterization is necessary.[1] The most fundamental component of modernity to come under scrutiny is the idea of "progress" and the assumptions contained within that powerful concept, especially that progress – of all sorts including material, intellectual, moral, and political – is inevitable and that through the application of human rationality (characterized as disembodied, instrumental reason) all problems can be solved. In our view, it is the acceptance of the idea of inevitable, linear progress that undergirds both the liberal democratic capitalist tradition as well as socialism–communism. Moreover, all the variations on those main currents that have taken the form of theories of modernization–development–dependency are fundamentally rooted in the commitment of modern political theorists to the proposition that things are and will become better, and that largely through the application of human reason through science and technology, the fundamental problem of scarcity will be resolved, or at least "rationally" managed.

If the research in biopolitics to date has revealed anything, it is the impoverishment of the belief in inevitable progress. Proponents of biopolitics have argued for the replacement of that idea with a view of evolutionary change that is informed by the life sciences, especially by the contemporary evolutionary theory, including the possibility of a theory of punctuated equilibrium that postulates abrupt and transforming change rather than gradualism (Corning, 1983; Masters, 1989; Schubert, 1989; Somit & Peterson, 1989). These theorists all realize that progress, however it may be defined, is relative to a particular set of environmental conditions. Moreover, the ongoing process of adaptation to changing environments, while (following Corning's use of synergy, holism, and upward and downward causality, 1983, 2011a, 2011b, 2012) tending to involve greater complexity, is inevitably contingent and, furthermore, that complexity at one level may lead to simplicity at another (see also Laszlo, 1991; Sutherland, 1973). Above all, these theorists recognize the interconnectivity among all levels of reality and the multicausality that characterizes living systems (see Chapter 2).

The prevailing ideologies of the West in the modern age other than conservatism (liberalism, socialism, and communism) have embraced to varying degrees a fundamental belief in the idea of progress. Conservatism, in its distinctive forms in various countries (e.g., traditional Islamic cultures), alone has resisted this progressive strain. However, adequately addressing the problems associated with a deteriorating natural environment necessitates the replacement of this idea of inevitable progress with a life-science-based knowledge of evolution and human behavior. Contemporary biopolitical theory must be built upon a life science approach that includes the big three E's – ethology, ecology, and evolution (following Schubert, 1989, p. 324). All three of these components are essential to our ability to understand man in nature.

Throughout this book, we have argued for a framework of political inquiry that rescues political science from a very particular set of problems – problems that primarily result from the desire to achieve maximum freedom for individuals, especially freedom to pursue their individual economic agendas, based on the erroneous assumption that this development is inherently progressive and that it inevitably contributes to the common good. From a biopolitical perspective, it is species survival that is the ultimate goal. A glimpse of what concerns us is reflected in William Connolly's suggestion that "... perhaps modernity is the epoch in which the destruction of the world followed the collective attempt to master it" (1988, p. 1). According to Connolly, our task is to "... call the modern project of mastery into question" (p. 1).

Again, we reiterate that biopolitics provides a better vantage point than other frameworks in contemporary political science to avoid the fundamental problems of what might be called hyperrationalism and hyperindividualism. Both represent extreme views closely associated with modernity in the West. Biopolitics can contribute to our avoidance of the hubris so often associated with modernism by reinforcing the need to see ourselves as part of a biological continuum rather than standing above and apart from other life forms. In stark contrast to the "possessive individualism" of liberalism (following MacPherson's, 1962, use of that expression) and the stifling collectivism often accompanied by the authoritarianism of socialism–communism in practice, ecopolitics seeks to promote the collective good by giving priority to the maintenance of a physical environment, thus increasing the likelihood of the survival of the human species. From a political and policy standpoint, more than anything else biopolitics challenges the priority of growth in general and economic growth in particular and of the protection of the extreme views of individual property rights as primary political and policy goals insofar as these goals begin to become threatening to our very survival. As noted in our discussion of Corning's work (1983) in Chapter 2, the policies that are needed are those that contribute to a successful survival strategy today.

Modern Rationalism and the Liberal Idea of Progress: Threats to Survival in the 21st Century

In 1929, Alfred North Whitehead delivered a series of lectures in which he offered the following as a definition of the function of reason: "The Function of Reason is to promote the art of life." Reason is then a tool or a process inherent in man as a result of evolution, which is utilized to sustain his existence through the modification of his natural, social, and cultural/symbolic world. As Whitehead so aptly describes the activity, the "higher forms of life are actively engaged in modifying their environment. In the case of mankind this active attack on the environment is the most prominent fact in his existence." Further, he categorizes the attack on the environment as a "... three-fold urge: (1) to live, (2) to live well, (3) to live better. In fact the art of life is first to be alive, second to be alive in a satisfactory way, and third to acquire an increase in satisfaction" (1929).

As a tool, process, or method, reason does not represent the complete armament of man in his quest for life; it is a tool, though certainly a powerful one. However, since the 18th century, reason, largely in the form of science and from science, technology, has come to dominate the range of responses made by man to his environment in the West. It has taken upon itself not only the function of helping man cope with his environment but also of passing judgment on the quality of that environment and radically altering that environment.

In politics, it has been argued that the use of reason has largely given way to the intellectual fashion described by the English political theorist Michael Oakeshott as modern rationalism. As Oakeshott sees it:

> The conduct of affairs, for the Rationalist, is a matter of solving problems, and in this no man can hope to be successful whose reason has become inflexible by surrender to habit or is clouded by the fumes of tradition. In this activity the character which the Rationalist claims for himself is the character of the engineer, whose mind (it is supposed) is controlled throughout by the appropriate technique and whose first step is to dismiss from his attention everything not directly related to his specific intentions. This assimilation of politics to engineering is, indeed, what may be called the myth of rationalist politics. And it is, of course, a recurring theme in the literature of Rationalism. The politics it inspires may be called the politics of the felt need; for the Rationalist, politics are always charged with the feeling of the moment .... How deeply the rationalist disposition of mind has invaded our political thought and practice is illustrated by the extent to which traditions of behavior have given place to ideologies, the extent to which the politics of destruction and creation have been substituted for the politics of repair, the consciously planned and deliberately executed being considered (for that reason) better than what has grown up and established itself unselfconsciously over a period of time. (1962, pp. 4, 21)

Thus, the rationalist, from this viewpoint, combines infatuation with method with an undying optimism, which leads him to believe that he can devise a solution for any problem confronting him. The only problem confronting the rationalist politician is the one of choosing the proper form of manipulation, a problem that is then solved with the identification of the dependent and independent variables and the elaboration of a causal model followed by the application of technology. As Simon (1983) has shown repeatedly, this view of rationality is a fiction and ignores the human behavioral tendency to "satisfice" or settle for something less than perfect rationality, but which is consistent with human nature. It also flies in the face of substantial research in cognitive neuroscience discussed in Chapter 5.

Such an approach to politics is clearly sustained by a faith in progress (Caton, 1990; Laszlo, 1991), the inevitability, and desirability of change. This is a Western attitude, according to Oakeshott, and is especially prevalent in the United States, which Oakeshott calls a classic case of rationalism in politics. Parenthetically, and in keeping with the biopolitical theory, one might ask whatever happened to the conservatism of Edmund Burke, who told us that "politics ought to be adjusted, not to human reasonings, but to human nature; of which the reason is but a part, and by no means the greatest part." The rationalist myth, which Burke struggled against in the theorizing of John Locke and as manifest also in the work of the liberal utilitarians, persists today in an only slightly different conceptual guise. In its most contemporary manifestation, this view undergirds the ideas of modernization and political development (see Chapter 4). The problem might be reduced to the problem of limiting the pursuit of individually defined goods when the collective good is endangered, following Garrett Hardin (1968), this results in the well-known problem of the "tragedy of the commons."

We are not suggesting that the idea of individual liberty or the expansion of human freedom is undesirable – absolutely not. Our view is that of Corning when he says that "survival is at once an individual and a collective affair – the two levels of adaptation are neither wholly distinct nor wholly concordant. To a considerable extent, the satisfaction of individual needs also furthers the interests of the collective survival enterprise. At the same time, transcendent 'species needs' may directly conflict with individual self-gratification" (1977, p. 62). The challenge, of course, is to be able to, politically speaking, assert and guarantee the interests of the collective survival enterprise when that is necessary. Still more important is the issue of what exactly constitutes the collective survival unit. This too is a moving target that must be understood within an ecological framework. Biopolitics reminds us that ultimately the unit is the species. Most recently, with the break up of the Communist bloc, the issue of what is the most important basis (unit) for a political order has thrust itself upon is in the form of virulent nationalistic and ethnic political movements that threaten the stability of numerous political orders, perhaps with good reason (Kaplan, 1996;

Moynihan, 1993; Pfaff, 1993). But, whether they will result in successful adaptations remains an empirical question that can only be answered over time. The recent tragedies in Rwanda and the former Yugoslavia, as well as the current struggle in the Ukraine, underscore this reality (see also Chapter 4 of this volume for a discussion of ethnic conflict and warfare).

ECOPOLITICS: A CHALLENGE FOR POLITICAL THEORY

Ecopolitics is a branch of biopolitics that focuses on the interplay of politics, economics, culture, and the environment in the processes of local, regional, national, and international policy making giving due consideration to the global ecosystem and its many subsystems and to the threats to those ecological systems. Glendon Schubert, early on in his important work *Evolutionary Politics* (1989), summarizes the gravity of the current situation for the human species. For Schubert, the success of humans in the competitive exclusion of other living species, particularly during the most recent 10,000 to 12,000 years since we began in a serious way to scarify the natural land with our agriculture, has by no means necessarily been adaptive for our species except from a point of view with as short a range as that. Our increasing technological capacity and tendency to eliminate other living species upon whom we previously had relied on for sustenance, as well as ourselves, is an

> index of the extent to which our trophic niche, defined as the functional status of an organism in its community, is being predetermined as much by the indirect as by the direct effects of our predatory activities; and the fossil record is replete with evidence of extinct species whose predation was so successful that they themselves starved to death. The issue goes far beyond the restoration of token vegetation to, and domestic pets from, urban areas or even the apparent trends in the direction of human conspecific predation ranging from licensed hunting in season (under circumstances such that other hunters present more frequent targets of opportunity than the crops of ruminants or rodents available for harvesting). It extends also to the ubiquitous predations now characteristic of all large urban areas in the United States (where only humans can be and are hunted by each other, at least in part because all other prey has been exterminated).... The human species cannot destroy the biological community of which humans have been a part without their degradation of that biological community returning as feedback to threaten the human political community .... (Schubert, 1989, pp. 17–18)

Numerous attempts have already been made (e.g., Boulding, 1978; Bryant & Bailey, 1997; Carter, 1999; Earley, 1997; Eckersley, 1998;

Hayward, 1998; Kassiola, 1990; Ophuls, 1977, 1997; Pirages, 1978; Woodhouse, 1972) to focus attention on the need for an ecological approach and to explicate the limitations of the liberal democratic capitalist model, particularly as that model becomes pursued on a global scale in the future. What these ecological critiques lack, for the most part, is a grounding in the biopolitical theory of the sort provided by Corning, Masters, and Schubert. Importantly, however, their ecological perspective could readily be accommodated within the biopolitical theory. The basic features of an ecological framework are by now fairly well known or are readily accessible to the interested reader. Ecology includes several key principles, which are sometimes presented as "laws" (e.g., see Kieffer, 1977; Odum, 1971; Ophuls, 1977).

The first of these principles is an ecosystem concept which states that everything is related to everything else. There are vital functional relationships between organisms and their environments. Subsystems of the environment are related to other subsystems and the whole must be understood holistically. Thus, there is widespread interdependence among living organisms and between them and their physical environments (see Corning, 2011a, 2011b; Laszlo, 1991). We are only now beginning to appreciate the real significance of these ecological principles as we face environmental problems of the sort presented by global warming or acid rain or the aftermath of a Chernobyl disaster. Yet another version of this first principle is the idea that you can never do just one thing. We cannot simply depend on imposing rigid parameters on our actions. This is precisely why we so often have to deal with the unanticipated consequences of our actions. Ecology also tells us to value diversity and complexity for these characteristics have been linked with stable ecosystems over time. And finally, ecology denies us the category of "waste" because everything goes somewhere. Each of these ecological insights presents specific challenges to the current conduct of politics and economics.

Accordingly, one of the most critical items on the agenda for biopolitics at this time, in our view, is the elaboration of the ecological component of a life science perspective for political theory and for politics. The relative absence of this dimension early on in biopolitics was duly noted by one of us (Hines, 1982). Among proponents of biopolitics, the work of Caldwell (1964, 1987, 1999) has been the exception. Political theory and public policy must be brought together to provide a new political paradigm (comparable to the liberal democratic capitalist paradigm of the 19th and 20th centuries) that will reflect the values necessary to promote ecosystem stability and sustainable development for the entire planet (see Funke, 2013).

In 1999, Lynton Caldwell, long an advocate for environmental issues, posed the question of whether the human species is destined to self-destruct because of our failure to understand the fundamental challenges presented to our survival as a result of the degradation of the environment. Indeed, following Eric Ashby, Caldwell accepts the idea that what we are faced with is not a crisis (which will pass) but a "climacteric" – an ongoing engagement by the human species with the problems of population, resources, and pollution. The full dimension of this climacteric is only now beginning to be realized. A brief review of the various components will have to suffice to make this important point here.

TRENDS OF THE CLIMACTERIC

Since 1984, The Worldwatch Institute has published an annual report, *State of the World*, as well as other publications that document the various threats we face. The 2013 and 2014 volumes in this series focus on the problem of sustainability. These publications continue to provide summaries of the various components of that challenge on an annual basis. Brown and coauthors (1999) identify seven trends that must be dealt with if we are to manage Caldwell's climacteric. These trends are population growth, rising temperature, falling water tables, shrinking cropland per person, collapsing fisheries, shrinking forests, and the loss of plant and animal species.

The steady increase in global population represents a formidable challenge indeed. In *Beyond Malthus*, Brown, Gardner, and Halweil (1999) identify 19 dimensions of the population challenge. As they note: "During the last half-century, world population has more than doubled, climbing from 2.5 billion in 1950 to 5.9 billion in 1998. Those of us born before 1950 are members of the first generation to witness a doubling of world population. Stated otherwise, there has been more growth in population since 1950 than during the 4 million preceding years since our early ancestors first stood upright" (1999, p. 17).

As staggering as this trend is, the prospects for continued population growth, despite declines in the industrial nations, are foreboding. United Nations' demographers project an increase over the next half-century of another 2.8 billion people. "In contrast to the last 50 years, however, all of the 2.8 billion will be added in the developing world, much of which is already densely populated" (Brown et al., 1999, p. 18). Clearly, this trend has profound implications for all of the other six trends mentioned above.

While population, per se, is not a problem directly for the developed world, the developing world continues to see the economic inequalities that characterize the relationship between the two worlds as a significant contributing factor to the population problems of the developing world. As of 2014, the U.S. Census Bureau estimates the world population at 7.168 billion and the United Nations predicts 10.9 billion by 2050 (see Fig. 1).

The most dramatic population growth in the future will come in Africa. Suffice it to say that the issue of population growth is engaged ideologically, politically, and economically on several levels. There is no consensus on how many people the Earth can support and on what should be done and who should do what (Cohen, 1995). Lomberg (2004) cover a wide range of global crises and propose various solutions to include climate change, communicable diseases, conflicts, access to education, financial instability, malnutrition and hunger, migration, sanitation, and access to clean water and others. Although space does not permit the discussion warranted by all of these pressing global issues, the challenge of water bears a few comments. In his major study of water and its availability in relation to historical struggles for wealth and power throughout human history, Solomon (2010) demonstrates that there are limits to natural carrying capacity in ecosystems and, therefore, constraints on human civilizations relative to the availability of those critical natural resources – air, water, arable land, and natural food supplies.

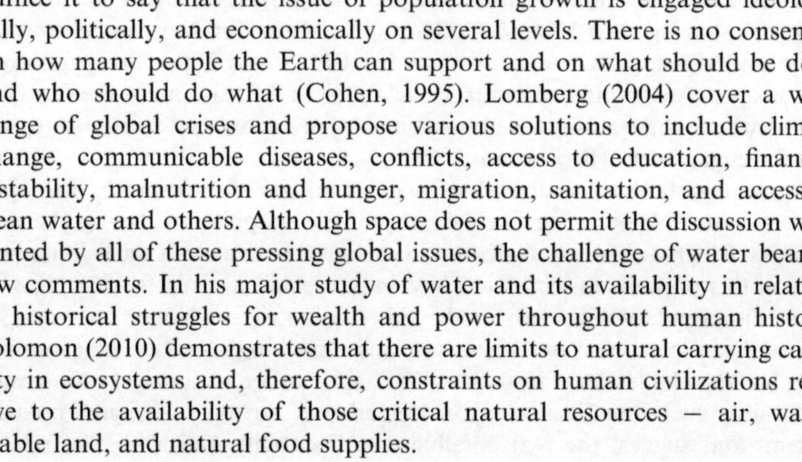

Fig. 1. World Population Growth, Actual and Predicted, 1800–2050. *Source*: UN International Programs Center (1995).

Similarly, Diamond in his work on the collapse of human societies has shown these critical limitations on human populations. Pirages and DeGeest (2004) have also stressed the case of freshwater and the oceanic commons. They note that "Access to water is a source of tension and potential conflict in many regions, since nearly half of the world's land area is in water basins shared by two or more countries. Nineteen of the world's largest rivers have five or more countries sharing the water basin. The Danube River has seventeen countries sharing its basin. The Congo and Niger rivers have eleven countries sharing their basins" (2004, p. 62). Moreover, the competition for water and other scarce resources has led China and India to aggressively pursue agreements and expand trade relations in Africa, one of the few continents where these resources abound and where economic relationships are ripe for engagement (Broadman, 2008).

It is important to realize that some countries, for instance, Ethiopia, Pakistan, and Nigeria, are likely to triple their population during the next 50 years. Clearly, these countries are ill prepared to deal with the full implications of this population increase. They, in particular, already feel the pressure from this population growth in the other areas of challenge, for example, cropland per person, water resources, and shrinking forests. The economic and health challenges faced in these and other countries are severe. Naturally, such conditions place tremendous stress on fragile political systems and suggest the real possibility of instability and a host of related problems. "Worn down by the struggle to deal with the consequences of rapid population growth, they are unable to respond to new threats, such as AIDS, aquifer depletion, and the flooding that can follow deforestation" (Brown et al., 1999, p. 112). Perhaps the greatest tragedy is that in some of these countries the result has been and will continue to be an increase in death rates. Whether this will solve the problem remains to be seen. In any event, it is hardly a desirable solution (see Bryant & Bailey, 1997).

As Caldwell (1987) observed some time ago, the full range of biopolicy encompasses the entire relationship between science and society, particularly those policy areas that are related to health and the environment. As he put it: "The building of a better bridge between science and society leads to consideration of four basic elements in the process. These are: first, prevailing perceptions of man's relation to nature; second, the meaning of science as interpreted by formalized education; third, communication between scientists and policymakers; and fourth leadership toward a synthesis of scientific knowledge and ethical values. Whatever utility, the concept of biopolitics possesses is primarily in relation to this fourth element" (Caldwell, 1987, pp. 7–8).

As proponents of the biopolitical theory and biopolicy construct a theoretical framework that is inclusive and links scientific inquiry, political philosophy and ethics, this new intellectual edifice can better support focused inquiry into biobehavioral research and biopolicy specializations that draw upon a wide range of disciplines. Political ecology needs the underpinnings of the biopolitical theory with its grounding in modern evolutionary biology and the life sciences if it is to realize its full potential. Because environmental policy issues are inevitably joined in a debate over the desire for economic growth and development and in the fulfillment of human needs and the collective goal of survival, it is essential that the discipline of economics experience a "greening" as well. To that end, after discussing the impassioned policy issue of climate change, we offer a brief examination of the case for an emerging interdiscipline of bioeconomics.

THE CLIMATE CHANGE/GLOBAL WARMING CONTROVERSY

From the perspective of ecopolitics and bioeconomics, the single most contentious issue in environmental politics and policy today is the one surrounding scientists' claim that global warming and climate change are the result of human action, specifically the reduction in the ozone layer resulting from carbon pollution. In 2014, the Intergovernmental Panel on Climate Change issued its latest and most comprehensive report. The full report is available at https://www.ipcc.ch/report/ar5/wg1/, while the Summary for Policymakers (SPM) is found at http://www.climatechange2013.org/images/report/WG1AR5_SPM_FINAL.pdf.

The panel's primary conclusion is: "Warming of the climate system is unequivocal, and since the 1950s, many of the observed changes are unprecedented over decades to millennia. The atmosphere and ocean have warmed, the amounts of snow and ice have diminished, sea level has risen, and the concentrations of greenhouse gases have increased" (SPM, 2014, p. 3). In addressing the drivers of climate change, the panel concludes: "Total radiative forcing is positive, and has led to an uptake of energy by the climate system. The largest contribution to total radiative forcing is caused by the increase in the atmospheric concentration of CO since 1750" (SPM, 2014, p. 13).

Despite the comprehensiveness of these reports and the abundance of evidence they provide, there is still resistance to their findings. As Funke

(2013, p. 145) observes, moneyed interests using powerful lobbying techniques have made extensive efforts to thwart attempts to establish policies in the United States that would restrain further production and consumption of oil and the carbon emissions that follow. The prospect of energy independence through the development of oil shale and Canadian oil sand supplies and the use of controversial technologies (e.g., fracking, see Chapter 6 for an in-depth analysis) that threaten the water table and the conversion of oil and natural gas to liquid nitrogen has emboldened those who advocate for greater independence from oil imports and herald an era where the United States could become an exporter of oil and become second to Russia in total oil production.

In contrast, a recent study by Archibald contends that advocates of global warming – which he calls a myth – based on the long view of climate change, has obscured the shorter solar cycle, which portends a very different challenge – that of global cooling that could result in drops in temperature in the United States of approximately 2.1 degree centigrade (2014, chapter 2). He argues that this will be accompanied with comparable drops in temperature in Europe and other parts of the world leading to reduced agricultural productivity and famine and drought that will create political unrest. He forecasts a 21st century in which life will be, in the terms of Thomas Hobbes, "nasty, brutish, and short."

No one has examined the political economy of oil more thoroughly than Daniel Yergin (2009, 2011). Together with Stanislaw (2002), he investigated the global competition to achieve a commanding position in the world economy. Yergin's latest book (2011) opens by relating the quake that rocked Japan and led to the worst nuclear accident since Chernobyl at the Fukushima Daiichi complex to the political turmoil in Tunisia, Libya, and then Egypt that dramatically increased the price of oil and raised fears of an energy crisis brought on by the possible expansion of conflict in and loss of oil from the Middle East. Yergin analyses the interrelatedness of population growth, increased consumer demand, security issues, and environmental hazards as the world economy struggles to sustain growth.

> The interaction of environmental concerns with energy will continue to shape the larger energy marketplace. The biggest question is climate change and carbon. Over 80 percent of world energy continues to be supplied by … carbon-based fuels. About 75 to 80 percent of world energy is expected to be carbon based two decades from now. The growing importance of the climate change question ensures that this ratio will be strongly challenged both politically and technologically as people strive to decarbonize energy. (Yergin, 2011, p. 713)

The most important linkage in both global warming and global cooling is the linkage between energy and the environment. Energy policies and environmental policies are inextricably bound up with one another as countries and international organizations struggle to maximize economic development and safeguard the natural environment. Together, energy and the environment have become as important as security issues. These security issues have been researched by a number of scholars (Barnett, 2001; Homer-Dixon, 1999; Pirages & DeGeest, 2004) and are the focus of the Millennium Project where research is undertaken to assess worldwide environmental-related issues in order to identify and analyze events that might trigger future international environmental treaties, conventions, or protocols and/or modifications to the existing ones. "The Millennium Project defines environmental security as environmental viability for life support, with three sub-elements: preventing or repairing military damage to the environment, preventing or responding to environmentally caused conflicts, and protecting the environment due to its inherent moral value" (http://www.millennium-project.org/millennium/env-scanning.html). As the science that supports global warming and climate change becomes even more persuasive, countries are beginning to act decisively to capture natural resources.

In the Arctic, for example, "[G]lobal warming has given birth to a new scramble for territory and resources among the five Arctic powers. Russia was the first to stake its claim in this great Arctic gold rush, in 2001. Moscow submitted a claim to the United Nations for 460,000 square miles of resource-rich Arctic waters, an area roughly the size of the states of California, Indiana, and Texas combined" (Borgerson, 2008). Although this claim was rejected, clearly the rush is on for oil, gas, and other resources, and the threats to the natural environment are real. Potential conflict over these claims, shipping lanes, and other issues will likely ensue given the lack of any "overarching political or legal structures that can provide for the orderly development of the region or mediate political disagreements over Arctic resources or sea-lanes" (Borgerson, 2008, p. 71).

THE RELATIONSHIP OF POLITICS AND ECONOMICS: THE IMPORTANCE OF BIOECONOMICS

Most postmodern approaches to the political theory have involved penetrating critiques of the foundations of knowledge in modern political theory, in particular the scientism–positivism of modernity with its attendant quest for certainty. Generally, they have focused upon metatheoretical issues rather than on practical politics. Or, if they have addressed practical politics, they have focused on various aspects of democratic politics or on the socioeconomic (social welfare) components of domestic public policy stressing problems of resource distribution. It seems generally to be the case that today's political scientists have not retained the comprehensiveness in their political theories that once characterized classical political philosophy and that today does characterize biopolitics.

Much of the original purpose of political philosophy, as reflected in the writings of Plato and Aristotle, has been cast aside in favor of the fragmentation of political philosophy into a large number of disciplines, including history, political science, anthropology, sociology, and economics. As Ophuls observes, "the perennial, but dormant, questions of political philosophy have been revived by ecological scarcity" (1977, p. 10; also see Sibley, 1972, 1977). From our point of view, the contemporary problems of the physical environment, such as global warming, desertification, deforestation, air and water pollution, and species extinction understood biopolitically (especially ecologically), cannot be addressed until there is widespread acceptance of the primacy of politics and a survival-oriented policy science over economics, or as Peter Corning recommends, the development of "bioeconomics" (Corning, 1997, 2003, 2005). That is, concern for the collective good (of societies and of the human race) must take precedence in political theory over the self-proclaimed needs of the individual.

Individual freedom is then, of necessity, to be limited if collective survival is to be guaranteed. Corning defines bioeconomics as "the study of how living organisms (of all kinds) acquire and utilize various resources to meet biological needs" (1997, p. 250; 2005, p. 241). Bioeconomics is emerging as an important interdiscipline that provides coherence for the search for and explanation of human biological, sociocultural, and political–economic adaptations to meet basic human needs (Corning, 2005: chapters 9 and 10). This bioeconomic pursuit of basic needs is depicted in Fig. 2.

Paramount among these needs is survival, and Corning (1997, p. 251) redefines economies as "survival enterprises." Clearly, such a view of economics and economies must include adaptation to the environment and reflect a concern for the stewardship of that environment over time in order to sustain the enterprise. Fundamentally, this view is in conflict with the prevailing ideology and sociocultural norms of the capitalist West. As Dietz and van der Straaten note: "Labour and capital will offer fierce resistance to the determination of ecological limiting conditions, especially when their short-term interests are unilaterally jeopardized. In fact, a social

Fig. 2. Basic Needs and Human Adaptation. *Source*: Adapted from Corning (2005, p. 284).

struggle concerning the distribution of the ecological space available among the various production and consumption processes will be the result" (1993, p. 139). Although they conclude that the state must play a central role in resolving these disagreements, such expansion of state control need not necessarily be incompatible with democratic political practices. However, there will be real problems finding common ground between liberal democratic capitalism and environmentalism, not to mention more radical environmental perspectives, for instance, ecologism, that may support authoritarian politics in the service of the environment – or not (Gat, 2007). This point relates back to our discussion of Somit and Peterson's examination of the role of human indoctrinability as a necessary, but not sufficient, condition for the possibility of democratic politics.

Interestingly, the origins of both the terms *economics* and *ecology* lie in their root concern with the management of the household, *oikos*. We believe, as does Corning, that what is needed now is a merger of biopolitics and bioeconomics to create biopolitical economies. The holism and contextualism of ecology needs to be a counterbalance to the extreme individualism of economic man as well as a narrowly understood sociobiological man. And the contextualism of ecology and evolution must be reflected in the epistemological foundations of science and social science (see Hines, 1979, 1982). A biopolitical economy should be based upon the original concept of a political theory that recognized the problematic features of the relationship of man to nature and to his fellow man. Because of the relevance to this issue, we quote at some length the political philosopher

Joseph Cropsey (1960) on this matter of the subsumption of economics to politics.

> Aristotle saw at the same time that acquisitiveness was "natural," that it was deplorable, and that it was indispensable to civil peace. What is most conspicuous in his discussion is that he conceived it to be deplorable. Aristotle deplored retail trade, usury, and "acquisitiveness" in the name of virtue, a quality of man which is perpetually threatened by his nature as that shows itself in his necessitousness. And yet his virtue and all his manifest possibilities are unintelligible except in the light of his nature. The public or political function of political philosophy is to turn attention toward the meaning of nature as end and to divert attention from nature as the beginning; or to turn attention toward nature as provident and friendly to human excellence, and away from nature as polemic and divisive, which it is. The classics seem to have believed that excessive emphasis upon man's neediness would blind him to the reason for not becoming a self-regarding atom.

Cropsey goes on to note, however, that the

> self-regarding man is, as such, the opposite of the citizen, whose peculiar virtue is patriotism or regard for country and countrymen first and foremost – say institutionalized altruism. Without that quality, men cannot take their proper place among the ruled and assume their proper posture with respect to the rulers. It is on this basis that the classical writers could simply comprehend the economic activity within or under the political, and discussion of the economic activities within the discussion of political life more generally. Successful political life depends upon a proper public emphasis upon those irenic elements of nature which are friendly to virtue and hence to political life, and the suppression or warding off of those polemic elements which are neutral or hostile toward the perfection of civil community. The instrument of that prudent emphasis is law, or convention, and its fruit is political society, which is an artifact, the product of provident men who are the great benefactors of their kind. They may be said to rise above nature through the breadth of their understanding of nature: perceiving that nature is in some ways friendly and in some ways indifferent to the perfection of political society, the classical writers declined to deduce political life from the simple laws of nature .... Political philosophy is the comprehensive social, not to say human, science, because it comprehends all the aspects of nature, the friendly, the indifferent, and even the hostile, under law that emanates from human discretion. It is based upon a narrow-eyed scrutiny of nature by discreet men who understood the precariousness with which political life is balanced upon its natural base. (Cropsey, 1960, pp. 10–11)

The fundamental purpose of a biopolitical economy is the management of the household. But the household of today is the planet. This framework invites the exploration of supranational and subnational regional political authorities as alternative political frameworks for dealing with environmental problems and priorities. Such a framework also depends upon the clear recognition of the political theory (and practice) as occupying the higher ground (i.e., being more comprehensive and prior to) rather than yielding that ground to the economic theory (and practice). The economic theory should be a derivative of the biopolitical theory, properly understood. At this stage in history, it is essential to reconsider the fundamental purpose of human politics in relation to the rest of the natural world. It is also necessary to appreciate the limitations of human "polytechnic rationality" (following Caton, 1990) as we seek to achieve our survival through politics.

Corning (2011a, 2011b) has examined the problem of creating a political order that is committed to meeting the basic needs of its members. In the *Fair Society*, he provides a trenchant critique of prevailing Western capitalist economic theory and the rational economic man model followed by a proposal for creating a new "biosocial contract" built upon the current understanding of human nature that has been expounded throughout this volume. In the West, the exaltation of rationality has proven to be problematic and the results of the resultant inequality and conflict over resources are only now being realized even though there have been warnings during and after the Enlightenment enshrined the power of human reason. Recently, the French economist, Thomas Piketty (2014) has offered a major critique of the current way in which capital is accumulated and distributed in the world. Based on his analysis of economic data from 20 countries, he argues that there is likely to be increased inequality that will stress to the limits our societies and pressure governments to respond in new ways. He cites the "ecological stimulus" as a potential source of change in the way the economic roots of inequality are addressed (Piketty, 2014, p. 568). His argument quickly generated a substantial response, particularly in the United States, both supportive and highly critical.

Similarly, Paul Kress suggests that "the most profound political (social and cultural) process of the 20th Century has been that which Max Weber called the 'rationalization of Western civilization'" (1983, p. 115). As Kress observes, Weber also used the more somber phrase *die Entzauberung der Welt*, or "disenchantment with the world." After noting the number of critics sympathetic to the warnings by Weber about rationalization, and advocating relentless criticism of contemporary politics, society, and culture as appropriate activities for political theorists, Kress offers the recommendation that we must confront "the alleged failure of liberalism" (1983, p. 119).

We recommend the framework of biopolitics as a vantage point from which to mount a serious critique of the fundamental flaw of classical liberal politics. The flaw is that liberalism has failed to provide the necessary political, social, and cultural restraining parameters to insure the primacy of the collective good, measured ultimately by our survival potential. The manifestations of that failure are present everywhere in the current attempts of advanced industrialized and postindustrialized nations to deal with the problems of economic recession, unemployment, health care, environmental degradation, and a host of related public policy issues. At a time when many in the West have been so ecstatic about the demise of Communist politics in Eastern Europe and the Soviet Union, it is essential that we do not ignore the real limitations of liberalism as an ideology of modernity and capitalism as an economic system whose viability may be in doubt in the longer run. In addition, it is necessary to examine critically the implications of a sociobiological theory of human politics given that it is more narrowly based on individual strategies for survival and incapable of sustaining a collective, cooperative basis for group (species) survival (see Bloom, 2000).

That there are profound limitations to human rationality, especially our attempts to rationally control the natural world seems only too obvious (Simon, 1983). What we need to understand, however, is how we have reached this position intellectually and historically. Toulmin's (1991) recent review of the historical origins of modernity reminds us that there were two separate strains of thought that ushered in the modern age. The first (16th century), which was soon eclipsed by the second (17th century), was a humanist strain associated with the early renaissance. The second, to which we have already referred, was modern scientific rationality, which Toulmin dates from approximately 1630. It is Descartes and Cartesianism that best exemplifies the second, and to date, the dominant strain in modernity (see Bluhm, 1984). Modern rationalism is absolutely decontextual. As a result, it stands in direct contrast to ecology and biopolitics, which are absolutely contextual and configurative (as recommended by Lasswell, 1968). The ahistorical nature of modern rationalism makes it incompatible with an ecological/evolutionary approach. It is a small wonder that the fundamental nature of postmodernist critiques has been an attack on rationality.

Within evolutionary biopolitics, the sociobiological and the systems-level research agendas compete while operating at different levels of analysis. Indeed, these approaches share a common heritage and seek explanations that are ultimate as well as proximate, but have difficulty agreeing on the claim that evolution can operate at the group or collective level. The work of Peter Corning, discussed above and in earlier chapters, exemplifies the view that synergism and cooperative relationships represent evolutionary trends that are as consistent with the modern evolutionary synthesis as are sociobiology and inclusiveness fitness. These claims can be explored through empirical research into the adaptive strategies that are adopted by individuals and groups as they struggle to survive and reproduce under conditions of scarcity and conflict, as was suggested in our discussion of political evolution in Chapter 4. Corning (1997, p. 249) describes a number of important theoretical and analytical linkages that have been forged between biology and economics over the past 20 years as documented by Ghiselin (1992) and Hodgson (1993). He also cites a larger body of work and notes the establishment of recently established journals, *Ecological Economics*, *Journal of Bioeconomics*, *Evolutionary Economics*, and others. If these efforts can move economists away from a closed system model to a living, open systems model that reflects the realities of ecology, then the hoped for interdiscipline may actually emerge, hopefully in conjunction with the further elaboration of the biopolitical theory.

CONCLUSIONS

This chapter has introduced a range of environmental policy issues that demand attention within biopolicy and point to a direction through which inquiry can be shifted to bring about bioeconomics and a robust biopolitical theory. Hopefully, others will find the challenge of developing a biopolitical alternative for the political theory that can address the climacteric compelling and that collectively we can succeed in moving beyond the limitations of modernity to a vision of a postmodern ecopolitics that is worthy of pursuit.

The climacteric presents an enormous challenge to our species from a survival standpoint. If we fail to appreciate the environmental impact of globalization – initially a popular but now an oft debated idea – we may not be able to live to regret it. Although there is evidence that we have at least begun to recognize the problem and that there are growing attempts to devise strategies for adapting ourselves to the task at hand (Hajer, 1995), movement to that end is slow. A key intellectual achievement in support of this effort will be the synthesis of the biopolitical theory with work on political ecology. As demonstrated here, these two important strands of contemporary political theory share much in common and, especially when combined with bioeconomics, can present a formidable theoretical perspective for the articulation of a life-science-informed biopolicy for the environment.

In advocating our approach to the study of the climacteric and all the dimensions of global ecopolitics, we are, nonetheless, rather pessimistic about the prospects of finding solutions to this Gordian knot.

The particular role of the United States in helping to find solutions hinges in no small part on America's ability to project its version of ecopolitics in its domestic and foreign policies. Whether the American version of liberal democratic capitalism, which has achieved so much, is capable of this task remains to be seen. Ultimately, from a biopolitical perspective much will turn on whether scientists and philosophers like Alexander (1987), Pinker (2011), Ridley (2010), and Tancredi (2005) and political and economic scholars like Ikenberry (2014), Kagan (2012), Kaplan (2012), and Sachs (2008) who find a biological basis for morality and altruism are correct in anticipating a more enlightened America as a force for the common good at home and abroad. If so, this could lead to a resurgent and enlightened West that achieves what Sachs (2008) chose as the title of his book – the *Common Wealth*. Hopefully, in the long run the knowledge attained by advances in evolutionary science joined with the biosocial sciences, rather than continue with politics as usual, will allow leaders of the world to make better choices for the global common good.

NOTE

1. Our characterization of modernity draws upon the following sources in particular: Toulmin, (1991), Bluhm (1984), Caton (1990), Connolly (1988), and Kress (1983).

CHAPTER 8

TOWARD HUMAN-CENTERED POLITICAL SCIENCE

Robert H. Blank

Over three decades ago, Thomas Wiegele in his seminal work on biopolitics concluded that political science must begin to consider the subjects of its investigation as biological beings with intellectual capabilities. Unfortunately, he was forced to write: "At the present state in the development of the discipline, we can characterize ourselves as having ignored an enormous amount of information about humanity and its real nature, that is its nature *as it is lived*" (1979a, p. 145). He goes on to state that because our vision has been narrowly focused down safe, well-traveled tunnels, we have been inattentive to the powerful findings of the life sciences. A decade later, Roger Masters concluded that "the social sciences remain largely untouched by research in the biological sciences" (1989, p. xii). As recently in 2004, Blank and Hines concluded that despite the significant work in biopolitics since 1990, in large part, these earlier assessments remain accurate.

What has changed since then, however, is that despite the continued inadequate attention paid to explanations of political phenomena through biological variables in political science per se, there has been considerable research activity on the topic. Still, within the context of other social sciences, particularly anthropology and psychology where evolutionary perspectives and biological grounding of behavior have become well

Politics and the Life Sciences: The State of the Discipline
Research in Biopolitics, Volume 12, 229–242

ISSN: 2042-9940/doi:10.1108/S2042-994020140000012007

established, as a discipline political science lags far behind. Only sociology appears to have been as unreceptive toward recognizing Darwinian ideas and integrating the life sciences into the mainstream of the discipline. The "fallacy of arbitrary exclusion" is still very much alive in political science. According to McManus, Laughlin, and d'Aquili, the "fallacy is committed whenever we, through ignorance or through adherence to a normative rule, exclude from consideration material efficiently present in the phenomenon being studied" (1979, p. 345). Although it might be an exaggeration to say that biological variables have been "arbitrarily excluded" by political scientists, mainstream political science has tended to dismiss them as unimportant. In part this might be because the methodologies and terminologies from the life sciences are foreign to the average political scientist and seen as not essential or at least not worth the effort to study. Although the inclusion of life science concepts and methodologies into political science is challenging and does complicate research, "if we aspire to precision and scientific credibility in our work," it will be worth the effort (Wiegele, 1979a, p. 2).

As argued here, in order to address these shortcomings, we must adopt a more comprehensive definition of human nature which includes in an operational way the biological dimensions as well as the natural and psychological dimensions. This requires a reevaluation of the basic assumptions that underpin the discipline. Ironically, this will move us back closer to the roots of our discipline of Merriam, Munro, and to visionary Harold Lasswell, who in his 1956 APSA presidential address urged a reintroduction of biological concepts and stressed the need to keep up to speed with new technologies in the life sciences (1968, also see Dryzek & Schlosberg, 1995, pp. 134–135). We agree with Master's assessment that the unwillingness to accept a biological perspective to human nature and politics is not likely to survive the onslaught of rapid advances in the life sciences. "The political process must sooner or later be fundamentally affected by the power to change not only the environment but also the behavior and genetic composition of humans themselves" (Masters, 1989, p. xii).

BIOLOGICAL DETERMINISM

One of the most often expressed claims of opponents of the more inclusive paradigm presented here is that it represents genetic or biological determinism or a form of social Darwinism. To some extent, this might still reflect the outrage against some of the early claims by proponents of sociobiology, although as Losco (2011) points out, the sociobiology debate has faded as did use of the term itself. However, largely this claim is misleading. Although it is essential to integrate a strong biological base to our study of politics and political behavior, this does not mean we must accept a biological determinism that "pays no heed to human purpose activities" (Wiegele, 1979a, p. 145). We have tried to demonstrate this in Chapter 2. Nothing in the research of political scientists in biopolitics reported here suggests this, and, in fact, most such scholars have gone to great lengths to dismiss biological determinism as contrary to fact. As noted by Segerstrale (1998), it is important to "explode the misconception" that biology is something that automatically precludes or excludes culture.

Although clearly there is a biological basis as to why and how we respond to environmental cues, the focus is on biological capacity not biological prescription and on possibilities not determination. The research in social and cultural neuroscience and epigenetics makes this unequivocal. We agree with Francis Fukuyama who states:

> No respectable biologist would deny that culture is important and often exercises an influence that can overwhelm natural instincts and drives. Culture itself – the ability to transmit behavioral rules across generations in a nongenetic way – is itself hard-wired in the brain and constitutes a major source of evolutionary advantage for the human species. But that cultural content rests on top of a natural substructure that limits and channels cultural creativity. (1999, p. 158)

What a biopolitical approach suggests, therefore, is not biological determinism, but rather a more balanced view of the interplay between nature and nurture in shaping political behavior. It simply declares that *both* biology and environment are critical and that neither should be excluded from consideration as the SSSM has done. Behavior continues to be affected by the environment because our genes, rather than programming behavior, program behavioral development and possibilities, thus obligating us to an interactive model of behavior as the type presented here. Rather than speak of biological determinism, we should speak of biological potentiality and ecological opportunity as well as environmental and ecological threats to our survival.

To be sure, there are many observers out there who misinterpret biological findings for their own ideological purposes, just as the zealous eugenicists did early in the 20th century. Their very presence, however, makes acceptance of an interactive paradigm even more crucial, for to deny or neglect the biological aspects of the human species in our models of reality makes everything we say easy to dismiss. It fuels the arguments of the determinists who can argue with some justification that political science is irrelevant because it ignores a major part of that reality, also essentially on ideological grounds. Thus, the denial of biological influences in the face of the mounting scientific evidence plays into the hands of the fanatics. The threat of biological determinism can be countered only by an interactive, balanced paradigm that accepts evidence of the biological influence on individual behavior, organizations, and government and that is not afraid to face the extremists on their own grounds armed with a good working knowledge of the life sciences. As noted throughout this book, there is simply too much evidence of the impact of biological variables on *human* behavior to dismiss their relevance for *political* behavior.

THE SUBFIELDS

Hopefully it is clear from the discussions in the previous chapters that biopolitics should not be treated as a separate subfield in political science that stands alone in isolation from the established subfields. Although the impact will vary, biological concepts and variables have much to contribute to every subfield, from the study of international relations and comparative politics to political theory and public administration. Given the wealth of understanding of human behavior emanating from the life sciences and its implications for the study of leadership, decision making, organizational behavior, and our very own conceptions of human nature, it has significance for public policy as well as public law and legislative and judicial behavior. To accept that biopolitics has significance in all subfields, of course, does not mean that it must always play a central role, but rather that to ignore its implications will result in the exclusion of potentially important variables. It may be that in some areas biological variables have limited explanatory value. It is possible, for instance, that health and nutrition are not the most important variables in political behavior, however, by excluding their consideration for political or other reasons means that we will be unable to make that determination one way or the other.

Political socialization is a notable example of the dangers of an environmental bias that has excluded biological variables. The conventional concept of political socialization views children as malleable and passive, existing in a patterned, one-dimensional process through which they are led through various agents to confirm to the standards of society. For White, these assumptions "effectively deny 'the animal' in us, that is, the genetic pre dispositions that we all begin life with and that may resist or even contradict environmental pressures" (1993, p. 1). The result is that, after huge public investments in extensive surveys of children in the 1960s and after 50 years of study, political socialization has failed to provide adequate cross-generational and cross-cultural explanations and as a subfield has lost much of its initial enthusiasm and promise. To assume that the political socialization process works in an evolutionary, genetic, and neurological vacuum fits the SSSM but it fails to satisfactorily explain behavior and, in fact, is often contradicted by work in other disciplines.

A BIOPOLITICAL APPROACH TO SURVIVAL

Although the development of empirically based, biopolitical theory at the individual and group levels remains a most difficult and demanding area where methodological issues continue to adversely affect the research agenda, biopolicy as a research area is prospering. Although much of the work in environmental and health policy is not cast in a formal evolutionary or biopolitical framework, it has clear connections, which must be elucidated. Linking evolutionary theory and strategies for survival to policy issues (e.g., population, energy, pollution, deforestation, climate change, and so forth) will provide greater justification for the salience of these issues and for their importance relative to policy issues that relate to economic growth and development and political development. The latter areas have tended to attract far more attention in the short run, but as pointed out in Chapter 7, it is the longer run that reveals how provisional and problematic many of these short-term growth strategies may turn out to be.

In a discipline that prides itself on being a science, prior exclusion of independent variables of proven importance in other behavioral sciences simply because they fail to fit our model, or are viewed as too difficult to research, is short sighted and potentially self-destructive. As Wahlke, Caldwell, Wiegele, and many others whose work has been reviewed here have claimed for many decades, we ignore or dismiss biological evidence at the risk of losing respectability as a science discipline and relevance in a fast changing world where sustainability and the survival and quality of life of our species cannot be taken for granted. In order to meet the demands of these threats to survival, we must adopt theories of political order and institutions that have been necessary for survival tactics throughout history and prehistory. Any model that refuses to even consider the evolutionary history of the human species in a changing environment is likely to fail. Research using an interactionist model of genes, body, consciousness, and environment understands the complex interrelationships among biological and environmental variables and excludes none from consideration. Allowable reductionist moves as well as emergent properties are fully integrated into this model of human behavior. A life sciences model also elevates the status of health and the environment, both of which are prerequisites as to what society and the good life represent. We need a healthy body politic *and* a healthy spaceship earth. As such, we must take very seriously the challenges that ecologists and environmental scientists warn us constitute severe and urgent threats to human survival.

THE CHALLENGES AHEAD

Even if the present upward trajectory of biopolitics is sustained, its position remains tenuous. Despite these compelling signs of life in biopolitics, the SSSM continues to be strongly entrenched and has proven itself to be quite resilient in the past. In addition to the ever-present ideological constraints, there are many practical issues that will moderate the growth of scientific biopolitics. Thus, many challenges face us that must be addressed, not only through more research but also through intensified and creative organizational activities. In other words, the problems a biopolitics' agenda faces remain substantial, making a transformation to biopolitical politics a worthwhile but difficult endeavor. Moreover, the strategies for meeting these challenges will, themselves, be controversial.

First, there is a natural division between different types of biopolitical approaches. This has been evident from the first efforts of the biopolitical movement that were divided into biobehavioral and evolutionary biopolitical theory camps, reflecting the micro- and macro-level emphases, respectively (see Blank & Hines, 2004). A proponent of biobehavioral biopolitics, Wiegele, for instance, although acknowledging the usefulness of ethology and evolutionary studies "for providing organizing concepts," pointed to the methodological problems involved in using the comparative perspective on animal behavior (1978). The bottom line is that while this diversity was beneficial from a developmental standpoint, the lack of a clear, coherent focus might have contributed to misperception and disparagement from those on the outside, even those who might have had been more sympathetic to a biopolitics' perspective.

This division is still present between research drawing on behavioral genetics and that based on evolutionary psychology. Although these approaches unquestionably overlap, they offer substantially different dimensions for analysis. A pure evolutionary approach focuses on species-typical characteristics, while genetics narrows this down in scope to smaller breeding groups and ultimately to the individual level with the interaction of genes and environments in the phenotype. In other words, while evolutionary approaches focus on the structure of universal mechanisms, genetic approaches focus on individual differences. While both approaches are legitimate and complementary, it is important to address and communicate the differences in order to avoid confusion and misunderstanding. Hence, it must be clear that biopolitics is not just one general approach but contains numerous subapproaches each designed to address distinct problems. While this division of labor needs to be externally communicated, it is at the same time important to integrate and balance the insights of both approaches within the field itself. How to accomplish this and how to interpret the existence of individual biological differences within more universality-oriented evolutionary perspectives is, of course, not something that the field of biopolitics needs to work out on in isolation since such questions pertain to every biologically informed discipline.

Second, as seen in Chapter 3, there are continuing methodological problems and gaps. Most political scientists have been trained in survey research methods and associated statistical techniques. Yet, while the reliance on verbal self-reports has generated many important insights, it does not provide us with ample access to the more deep-seated processes often involved in biology-grounded responses. For example, as discussed in Chapter 5, neuroscience research demonstrates that much human action is unconscious (or at least not fully conscious) and that thus we are unlikely to be able to verbalize much of our behavior. This is especially the case for deep-seated and highly emotional responses such as fear, hatred, and aggression. Moreover, motor responses from the nervous system that react to cognitive and affective ideation act simultaneously with activity of the autonomic nervous system and the glandular and hormonal systems, and, thus, much of our emotional state and our physical readiness to act are a direct expression of the neurophysiological system. Therefore, although words express ideas overtly, they are embedded in a continual interplay of neural, hormonal, biochemical, and other physiological activity that make them at less-than-optimal approximations of deeper impulses. "Because behavior is often driven by automatic mechanisms, self-reports of mental processes are notoriously unreliable and susceptible to many forms of contamination" (Lieberman, Schreiber, & Ochsner, 2003, p. 682).

Early proponents of biobehavioral research began to apply methods taken from ethology, primatology, and psychology. Path-setting empirical research in biopolitics by Barner-Barry (1996), Salter (1996), Masters and Way (1996), Schubert (1996), and Kitchen and Peterson (1996) demonstrated the applicability of methods derived from the life sciences for the study of political behavior in natural settings. For example, while the functioning of motor responses is inaccessible to conscious, introspective analysis, they can be observed through a wide array of psychophysiological activity, such as sweating palms, release of adrenalin into the blood stream, rise or fall in heart rate, increase or decrease in blood pressure, slower or more rapid respiration, as well as through nonintrusive neural imaging techniques. One challenge in this respect is that the appreciation of and the ability to engage with biopolitical research depends on an understanding of such techniques which most political scientists lack. The success of biopolitical research within the political science community will, in other words, not just depend on the strength of its methodological choices but also on its ability to communicate and explain these choices to others who have not had formal training in the specific methods.

Third, researchers in the field of biopolitics (at least, those from political science) also face significant challenges as they will need continual retraining to keep up with methodological developments in the life sciences, not just to use these methods themselves, but also to fully understand other relevant studies. As a consequence, biopolitical research is best carried out as collaborative endeavors with researchers from numerous disciplines. As with any collective action, research collaborations become increasingly difficult to manage as the number of participants expands, thus adding a whole new set of obstacles to the conduct of biopolitical research (see Hatemi & McDermott, 2010).

Until now, most of the work in biopolitics has been conducted by persons trained in political science who have, primarily on their own, become familiar with the life science side of things. One of the results of the lack of formal training in biopolitics and the inadequate recognition of such work in mainstream political science is that young faculty who hope to receive tenure often have had to forego their interest in the life science contributions and publish in more traditional areas. Although they might publish excellent work with a biopolitics focus, it is often sporadic. The result is that there have been short bursts of activity in specific areas of biopolitics research, often followed by long gaps as that person's involvement subsides and before others take up the slack. This unfortunate pattern is evident in many policy- and behavior-oriented areas outlined here, for instance, the impact of health and age on behavior. This problem can be rectified only by a combination of recruitment of more researchers into politics and the life sciences and by the paradigm shift argued for in this book. Although it is unrealistic to expect political scientists to obtain advanced degrees in biology, they need to learn enough to converse appropriately with the hard scientists on interdisciplinary teams (Alford & Hibbing, 2008).

McDermott claims that the current research climate provides an "auspicious opportunity to undertake foundational investigations at the intersection of the natural and social sciences to produce transformative work with broad import for society" (2011a, p. 98). However, while there has been abundant work examining the genetic, neurobiological, and neuropsychological bases of social and political behavior, much of this work has been conducted independently in different disciplines, resulting in considerable duplication and a lack of a coherent core of social and political models and theories to guide it. More synchronization across fields is essential for methodical investigation and to provide additional gains in employing the skills, methods, and models of the natural sciences to address some of the most endemic social and political problems. The intersection between genes, cognition, culture, and behavior remain myriad, complex, and largely unknown, but increasingly susceptible to systematic investigation (McDermott, 2011a, 2011b).

While social scientists have tended to dismiss or ignore the importance of biological factors such as the brain and genes in their analyses, correspondingly, life scientists often have restricted their investigations to their own, often narrow, areas without applying their methods or findings to the wider social world. Researchers working in both areas need to merge their skills to investigate the nature and dynamics of large-scale social processes using methods and models based on individual biology, cognition, and genetics. To this end, Alford and Hibbing (2008) urge political scientists to join interdisciplinary teams so that nature and nurture can both be employed to understand the sources of mass-scale human politics. This exhortation echoes the call for interdisciplinary teams of researchers by Wiegele in his 1979 book, *Biopolitics*. Hall (2009), too, reiterates the need for interdisciplinary research.

According to a National Academy of Sciences' committee, interdisciplinary research is a mode of inquiry that "integrates information, data, techniques, tools, perspectives, concepts, and/or theories from two or more disciplines or bodies of specialized knowledge to advance fundamental understanding or to solve problems whose solutions are beyond the scope of a single discipline or area of research practice" (Committee on Facilitating Interdisciplinary Research, 2005). Thus, integration goes beyond merely appropriating ideas from other disciplines and demands the assimilation of previously unconnected elements, thereby generating novel theories, concepts, or methods that draw on and are of relevance to both disciplines, rather than having one act as the donor and the other as a recipient. Although political science has always espoused interdisciplinary research, Druckman, Kuklinski, and Sigelman found little evidence of true interdisciplinary work in political psychology, but rather "misused concepts and scant evidence of conceptual or disciplinary integration" (2009, p. 503). Their study also highlights the difficulty of achieving successful interdisciplinary research. The unity of science thesis assumes that scholarship ranges from "higher" (less fundamental) to "lower" (more fundamental) levels of analysis: thus, from political science to psychology, from psychology to biology, from biology to chemistry, and from chemistry to physics (Wilson, 1998). Under this assumption, appropriation from below is far more prevalent than interdisciplinary research among disciplines of more or less equal standing. For instance, political scientists would borrow heavily from psychology, but only rarely would psychologists borrow from political science.

Druckman et al. (2009) content analysis bears this out by establishing that political scientists are distinctively net importers of findings from other disciplines including psychology. This means that the conclusions in political science seldom reach other disciplines. One exception is Alford, Funk, and Hibbing (2005) who enjoyed considerable attention in other disciplines, even those outside the social sciences. Their findings have generated not just attention from within political science but from the life sciences as well (Blank & Petersen, 2013).

Effective interdisciplinary work, however, has been severely constrained by inadequate interaction and engagement by members of each field in the work of the other. Political scientists have borrowed from psychology, directly at first and then progressively indirectly, with the resulting conceptual usages evolving in ways that often bear little resemblance to their origins and evolution in psychology. "Although solo scholars from one discipline might immerse themselves in another discipline, one scholar knowing two disciplines does not equal scholars from two or more disciplines working jointly" (Druckman et al., 2009, p. 12). The key to successful interdisciplinary research lies in teamwork: members of different disciplines working on a single project. However, several preconditions are needed to make productive interdisciplinary exchange possible in biopolitics. First, political scientists must immerse themselves even more deeply in the theories, concepts, and methods of the life sciences. Furthermore, they must contemplate more fully what they can contribute to such collaborations by more carefully defining the key elements of political contexts.

Second, such forms of collaboration are unlikely without considerable institutional investment in fostering interdisciplinary work. Specifically, colleges and universities need to provide ample funding for such interdisciplinary research and weigh more heavily such activity in hiring and tenure and promotion decisions. Moreover, funding organizations, professional societies, and journals must solicit and support interdisciplinary research by ensuring appropriate evaluative standards and outlets (e.g., interdisciplinary grants, special journal sections on interdisciplinary research). Currently, these types of institutional incentives and opportunities in political science, although not unheard of, are quite rare. Even so, political scientists and social scientists in general must address more explicitly the extent to which they value interdisciplinary endeavors and, if they do, then the institutions governing research practices need to be brought more closely in line with the goals of interdisciplinary research.

For example, much of the published work in political neuroscience has been conducted by scholars in other fields and published in journals outside of political science. Although the work is valuable, as noted by Theodoridis and Nelson (2012), those pursuing it are not usually concerned with the same questions motivating those whose primary interest is politics. Certainly, the active participation of political behavior scholars might help make the results of such studies more germane to political science. At a practical level, this venture would benefit from greater acceptance of coauthorship such as is happening in political psychology. However, papers with multiple authors are not as much the norm in political science as they are in psychology and neuroscience. Excessive discounting of professional credit for scholarship produced in collaboration with many authors will continue to discourage work in political genetics and neuroscience. The recognition that this sort of research is far more likely to be done in collaborative settings increases the likelihood that political scientists will be involved in collaboration with psychologists and cognitive scientists, even as a third or fourth author.

The need for true interdisciplinary approaches opens a fourth challenge – that of institutional acceptance and support of interdisciplinary research. The extent to which, for example, PhD students are able to spend time familiarizing themselves with methods and theories from other disciplines are fully contingent on support from their department. One way to facilitate this support from organizations such as APLS is to set up summer institutes or short courses specifically directed at PhD students and young scholars with interest in biopolitics. Similar institutes in bioethics at the University of Virginia and elsewhere have been very successful in generating long-term interests in bioethical issues by scholars selected from a range of academic areas. Moreover, anything that can be done to encourage young political scientists to attend interdisciplinary efforts such as the 2010 Institute on Cultural Neuroscience and Biology of Social Cognition or the 2009 Illinois Biology and Politics Summer Institute on Sociogenomics will be useful in that regard. Alford and Hibbing (2008) caution, however, that while a critical mass of political scientists possessing a working knowledge of biology is necessary, it is not sufficient, since data availability will still be a concern.

The vast majority of political scientists lack the funding sources and research opportunities to take part in such research, thus creating a situation in which only those in the life sciences, with the methods and samples, but with less knowledge of the phenotypic properties of many social traits, are producing most of the research (Jasny, Kelner, & Pennisi, 2008). As discussed in Chapter 5, many factors that are of considerable interest to political scientists, such as political attitudes and behavior, have been widely studied by neuroscientists, behavioral geneticists, and others. Therefore, if social scientists cannot do their own research using these tools, a good part of the future and most novel research in their fields will be ceded to those not in the social science disciplines, or at the very least to only a small number of social scientists who have a foot in both worlds, meaning that those with the most knowledge of the social traits, context, and history will be likely selected out of the study of, design, and eventual scholarship on these traits. Although neurologists are much more adept at neuroimaging than political scientists, they are presumably more deficient in the context of the traits of interest to the social sciences (Hatemi, 2011).

In addition to the disparities in knowledge and training between the life and social sciences, the epistemological divide is buttressed by conflicting approaches to the scientific method, according to Hatemi and McDermott (2012b). They note that while biological science relies on the aggregation of knowledge and incremental discovery, the social sciences rarely experience rapid technological advances or undergo periods of fundamental discovery. Therefore, although no single approach or method is dominant or fixed in the life sciences, debates in political science tend to center on a single approach or method for all social research. Thus, any defect or shortcoming of a particular approach is used to disparage the entire research agenda, oftentimes for ideological reasons such as wholesale hostilities to biological work because of fear of past abuse or threats to currently accepted paradigms (e.g., Beckwith & Morris, 2008). Ultimately, these often unstated divisions present significant obstacles to the establishment of a common language, the means of evaluating research and the understanding of the methods and limitations (Hatemi & McDermott, 2012b).

CONCLUSION: LOOKING TOWARD THE FUTURE

While the creation of the APLS and *Journal of PLS* and the decision of the first generation of biopolitics proponents to "go it alone," at least retrospectively, could be regarded as a miscalculation, they have proven to be valuable forums for exchanging ideas and producing an impressive literature within an accommodating environment. The Association and the Journal have also provided a congenial context for personal contacts with like-minded scholars and the fostering interdisciplinary research within political science. In fact from its roster of members as well as those who published in its journal, the APLS is one of the most diverse academic bodies in terms of discipline and country. Moreover, the APLS has also served, in effect, as an extended support group for academics who have been spurned by the disciplines in which they were trained. Despite this, there needs to be more effort to publish in mainstream political science journals, which as evidenced above are becoming more amenable, as well as in leading science and biology journals such as the article in *Science* by Fowler and Schreiber (2008). Furthermore, ideas such as the Founders' project and Darren Schreiber's list serve in neuropolitics as well as APLS list serve are excellent mechanisms for publicizing the notable work being done in biopolitics even though most current members are already among its proponents.

In summary, although political science until recently has not been receptive to the arguments advanced by proponents of biopolitics, this resistance is weakening in part due to the groundwork laid by the early generation of biopolitics scholars, but mainly because of changing circumstances. The evidence emerging from the biological sciences, especially behavioral genetics and cognitive neuroscience, makes it increasingly improbable that informed observers can continue to dismiss biological concepts, variables, and methodologies in the study of political phenomena. Likewise, new research techniques, especially noninvasive brain imaging and candidate gene association studies, provide an expanding array of innovative opportunities for studying human behavior. Furthermore, the established field of evolutionary psychology provides a critical connecting link between the biological bases of behavior and the larger process of evolution in the study of political, social, cultural, and economic behavior (e.g., social scientists other than political scientists who are advocates of biosocial science, see de Waal, 1996; Walsh, 1995; Wright, 1994).

As serious scholarly discussion gradually displaces the more inflammatory ideological rhetoric that characterized the initial reaction to sociobiology and biopolitics, it may be that the inclusion of biological variables in greater numbers in the study of political behavior or the incorporation of insights from evolutionary biology will come to be viewed as essential, even welcome, though not necessarily the norm. In the interim – and in order to further this process – what is needed is empirically based research utilizing the most recent knowledge and techniques from the life sciences focused on problems of interest to the general political science audience and the new generation of students.

Gary Johnson (2011) views biopolitics as an unfinished revolution still in its early stages and argues that this revolution is significantly more fundamental than most of the scientific breakthroughs often referred to as revolutionary. Rather than simply an important new approach or theory, it involves a fundamental overhaul of the explanatory paradigm that humans appear always to have used. Just as the Copernican and Darwinian revolutions took over a century to succeed, the as-yet-unnamed revolution is the most profound of these three great revolutions because it assumes that the causes of human behavior are not fundamentally different than the causes of behavior of other species. Even when it was no longer possible to accept the Earth as the center of the universe, or the human body as a distinctive creation, it was still possible to believe that human behavior and culture, with their own governing laws fully independent of human biology, are the equivalents of special creations. This revolution, therefore, will prove the most profound and provocative because it assumes the first two and then proceeds to remove the last excuse for treating humans as if they are a thing apart from the rest of nature. Moreover, many who accept the second revolution reject the third. While there are sure to be some advances that lead to surges of progress, the revolution will largely proceed incrementally – article by article, book by book, course by course, and conference by conference. Hopefully, this book has contributed in a small way to this developing transformation.

REFERENCES

Climate Connections. (2012). Ohio: Fracking brine gas-well waste full of radium. *Climate Connections*. Retrieved from http://climate-connections.org/2012/09/04/ohio-fracking-brine-gas-well-waste-full-of-radium/. Accessed on March 9.

Press Connects. (2012). Gas field workers cited in PA hospital's losses. *Pressconnects.com*. Retrieved from http://www.pressconnects.com/viewart/20121224/NEWS11/312240043/Gas-field-workers-cited-Pa-hospital-s-losses. Accessed on December 24.

Abeles, R. B., & Ory, M. G. (1991). Aging, health and behavior. *ICPSR Bulletin*, 1–5.

Abrams, H. L. (1999). Can the twenty-fifth amendment deal with a disabled president? Preventing future White House cover-ups. *Presidential Studies Quarterly*, *29*(1), 115–133.

Adkins, R. E., & Dowdle, A. J. (2002). The money primary: What influences the outcome of pre-primary presidential nomination fundraising? *Presidential Studies Quarterly*, *32*(2), 256–275.

Adolphs, R. (2010). What does the amygdala contribute to social cognition? *Annals of the New York Academy of Sciences*, *1191*, 42–61.

Adorno, T. W., Frenkel-Brunswik, E., Levinson, D. J., & Sanford, R. N. (1950). *The authoritarian personality: A volume in the studies of prejudice*. New York, NY: Harper.

Alexander, R. D. (1979). *Darwinism and human affairs*. Seattle, WA: Washington University Press.

Alexander, R. D. (1987). *The biology of moral systems*. New York, NY: Aldine De Gruyter.

Alford, J. R., & Hibbing, J. R. (2008). The new empirical biopolitics. *Annual Review of Political Science*, *11*, 183–203.

Alford, J. R., Funk, C., & Hibbing, J. R. (2005). Are political orientations genetically transmitted? *American Political Science Review*, *99*, 153–167.

Allik, J., & McCrae, R. R. (2004). Toward a geography of personality traits: Patterns and profiles across 36 cultures. *Journal of Cross-Cultural Psychology*, *35*(1), 13–28.

Almond, G. (1990). *A discipline divided: Schools and sects in political science*. Beverly Hills, CA: Sage Publications.

Altman, L. K. (2007). TB patient is isolated after taking two flights. *New York Times*. Retrieved from http://www.nytimes.com/2007/05/30/us/30tb.html

Alvarez, R. A., Pacala, S. W., Winebrake, J. J., Chameides, W. L., Steven, P., & Hamburg, S. P. (2012). Greater focus needed on methane leakage from natural gas infrastructure. *PNAS*, *109*(17), 6435–6440. Retrieved from www.pnas.org/cgi/doi/10.1073/pnas.1202407109

Amodio, D. M., Jost, J. T., Master, S. L., & Yee, C. M. (2008). Neurocognitive correlates of liberalism and conservatism. *Nature Neuroscience*, *10*, 1246–1247.

Antfolk, J., Lindqvist, H., Albrecht, A., & Santtila, P. (2014). Self-reported availability of kinship cues during childhood is associated with kin-directed behavior to parents in adulthood. *Evolutionary Psychology*, *12*(1), 148–166.

Apicella, C. L., & Cesarini, D. A. (2011). Experimental evidence from the 2008 presidential election. In P. K. Hatemi & R. McDermott (Eds.), *Man is by nature a political animal: Evolution, biology, and politics* (pp. 261–272). Chicago, IL: University of Chicago Press.

App, B., McIntosh, D. N., Reed, C. L., & Hertenstein, M. J. (2011). Nonverbal channel use in communication of emotion: How may depend on why. *Emotion, 11*(3), 603.

Apter, D. (1996). Comparative politics, old and new. In R. Goodin & H. D. Klingemann (Eds.), *A new handbook of political science* (pp. 342–400). Oxford: Oxford University Press.

Arceneaux, K., Johnson, M., & Maes, H. H. (2012). The genetic basis of political sophistication. *Twin Research and Human Genetics, 15*, 34–41.

Archibald, R. (2014). *Twilight of abundance: Why life in the 21st century will be nasty, brutish, and short.* Washington, DC: Regnery Publishing.

Arnhart, L. (1988). Aristotle's biopolitics: A defense of biological teleology against biological nihilism. *Politics and the Life Sciences, 6*, 173–229.

Arnhart, L. (1992). Feminism, primatology, and ethical naturalism. *Politics and the Life Sciences, 11*, 157–178.

Arnhart, L. (1994). The Darwinian biology of Aristotle's political animals. *American Journal of Political Science, 38*, 464–485.

Arnhart, L. (1995). The new Darwinian naturalism in political theory. *American Political Science Review, 89*, 389–400.

Arnhart, L. (1998). *Darwinian natural right: The biological ethics of human nature*. Albany, NY: State University of New York Press.

Arnhart, L. (2005). *Darwinian conservatism*. Charlottesville, VA: Imprint Academic.

Arnhart, L. (2010). Biopolitical science. *Politics and the Life Sciences, 29*(1), 24–47.

Arrow, K. J. (1974). *The limits of organization*. New York, NY: W. W. Norton and Company.

Axelrod, R. (1984). *The evolution of cooperation*. New York, NY: Basic Books.

Balachandran, N., & Glass, D. J. (2012). PsychTable.org: The taxonomy of human evolved psychological adaptations. *Evolution: Education and Outreach, 5*(2), 312–320.

Balachandran, N. (2011). A proposed taxonomy of human evolved psychological adaptations. *Journal of Social Evolutionary and Cultural Psychology, 5*, 194–207.

Ballew, C. C., & Todorov, A. (2007). Predicting political elections from rapid and unreflective face judgments. *Proceedings of the National Academy of Sciences, 104*, 17948–17953.

Barash, D. P. (1977). *Sociobiology and behavior*. New York, NY: Elsevier.

Barkow, J. H., Cosmides, L., & Tooby, J. (Eds.). (1992). *The adapted mind: Evolutionary psychology and the generation of culture*. New York, NY: Oxford.

Barner-Barry, C. (1996). Nonparticipant observational research techniques. In A. Somit & S. Peterson (Eds.), *Research in biopolitics* (Vol. 4, pp. 1–22). London: JAI Press.

Barner-Berry, C. (1977). An observational study of authority in a pre-school peer group. *Political Methodology, 4*, 415–459.

Barner-Berry, C. (1978). The biological correlates of power: Dominance and attention structure. Paper presented at annual meeting of the American Political Science Association, New York, NY, 29 August.

Barnett, J. (2001). *The meaning of environmental security: Ecological politics and policy in the new security era.* London, UK: Zed Books.

Baron-Cohen, S. (2010). Empathizing, systemizing and the extreme male brain theory of autism. *Progress in Brain Research, 186*, 167–175.

Baron-Cohen, S. (2011). *Zero degrees of empathy: A new theory of human cruelty*. New York, NY: Allen Lane/Basic Books.

Barrett, L., Dunbar, R., & Lycett, J. (2002). *Human evolutionary psychology*. Princeton, NJ: Princeton University Press.

Bartels, L. (2013). *Your genes influence your political views. So what?* Retrieved from http://www.washingtonpost.com/blogs/monkey-cage/wp/2013/11/12/your-genes-influence-your-political-views-so-what/. Accessed on February 14, 2014.

Baumgartner, T., Heinrichs, M., Vonlanthen, A., Fischbacher, U., & Fehr, E. (2008). Oxytocin shapes the neural circuitry of trust and trust adaptation in humans. *Neuron, 58*, 639–650.

Bechara, A., Damasio, H., Tranel, D., & Damasio, A. R. (1997). Deciding advantageously before knowing the advantageous strategy. *Science, 275*(5304), 1293.

Beckwith, J., & Morris, C. A. (2008). Twin studies of political behavior: Untenable assumptions? *Perspectives on Politics, 6*(4), 785–791.

Berns, G. S., Blaine, K., Prietula, M. J., & Pye, B. E. (2013). Short-and long-term effects of a novel on connectivity in the brain. *Brain Connectivity, 3*(6), 590–600.

Bethlehem, R. A., Baron-Cohen, S., van Honk, J., Auyeung, B., & Bos, P. A. (2014). The oxytocin paradox. *Frontiers in Behavioral Neuroscience, 8*, 48.

Binder, L. (1971). *Crises and sequences in political development*. Princeton, NJ: Princeton University Press.

Blais, A., & Vincent, S. L. St. (2011). Personality traits, political attitudes and the propensity to vote. *European Journal of Political Research, 50*, 395–417.

Blanchard, K. C. (2009). The Illinois biology and politics summer institute. *Politics and the Life Sciences, 28*(2), 90–93.

Blank, R. H., & Hines, S. (2004). *Biology and political science*. London: Routledge.

Blank, R. H., & Petersen, M. B. (2013). Success or failure? A critical but optimistic evaluation of biopolitics. In S. A. Peterson & A. Somit (Eds.), *The world of biology and politics: Organization and research areas* (Vol. 11, pp. 199–216). Research in Biopolitics. Bingley, UK: Emerald Group Publishing Limited.

Blank, R. H. (1981). *The political implications of human genetic technology*. Boulder, CO: Westview Press.

Blank, R. H. (1982). Biopolicy: A restatement of its role in politics and the life sciences. *Politics and the Life Sciences, 1*(1), 38–51.

Blank, R. H. (1990). *Regulating reproduction*. New York, NY: Columbia University Press.

Blank, R. H. (1999). *Brain policy*. Washington, DC: Georgetown University Press.

Blank, R. H. (2006). The brain, aggression and public policy. *Politics and the Life Sciences, 24*(1–2), 12–21.

Blank, R. H. (2011). Biopolicy after three decades. *Politics and the Life Sciences, 30*(1), 52–55.

Blank, R. H. (2013). *Intervention in the brain: Politics, policy, and ethics*. Cambridge, MA: MIT Press.

Blank, R. H., Caldwell, L. K., Wiegele, T. C., & Zilinskas, R. A. (1988). Toward better education in biopolitics. *Issues in Science and Technology, 4*(3), 51–54.

Bloom, H. (2000). *Global brain: The evolution of mass mind form the big bang to the 21st Century*. New York, NY: Wiley.

Bluhm, W. T. (1984). *Force or freedom: The paradox in modern political thought*. New Haven, CT: Yale University Press.

Boix, C., & Rosenbluth, F. (2014). Bones of contention: The political economy of height inequality. *American Political Science Review*, *108*(1), 1–22.

Bonnicksen, A. L. (2013). The biopolitics program at Northern Illinois University. In S. A. Peterson & A. Somit (Eds.), *The world of biology and politics: Organization and research areas* (Vol. 11, pp. 34–37). Research in Biopolitics. Bingley, UK: Emerald Group Publishing Limited.

Booth, A., & Welsh, S. (1976). Stress, health and political participation. Paper presented at International Political Science Association Meeting, Edinburgh, Scotland.

Borg, C., & de Jong, P. J. (2012). Feelings of disgust and disgust-induced avoidance weaken following induced sexual arousal in women. PloS One, 7(9), e44111.

Borgerson, S. G. (2008). Arctic meltdown. *Foreign Affairs*, *87*(2), 63–77.

Borkenau, P. (1992). Implicit personality theory and the five-factor model. *Journal of Personality*, *60*(2), 295–327.

Borkenau, P., Mauer, N., Riemann, R., Spinath, F. M., & Angleitner, A. (2004). Thin slices of behavior as cues of personality and intelligence. *Journal of Personality and Social Psychology*, *86*(4), 599.

Boulding, K. (1978). *Ecodynamics: A new theory of social evolution*. Beverly Hills, CA: Sage Publications.

Boyd, R., & Richardson, P. J. (1980). Sociobiology, culture and economic theory. *Journal of Economic Behavior and Organization*, *1*(2), 97–121.

Braibanti, R. (1969). External inducement of political-administrative development: An institutional strategy. In R. Braibanti (Ed.), *Political and administrative development*. Durham, NC: Duke University Press.

Braibanti, R. (1976). Context, cause and change. In J. Hallowell (Ed.), *Prospects for constitutional democracy* (pp. 165–182). Durham, NC: Duke University Press.

Broadman, H. G. (2008). China and India go to Africa. *Foreign Affairs*, *87*(2), 95–109.

Brown, L., Gardner, G., & Halweil, B. (1999). *Beyond Malthus*. New York, NY: W. W. Norton.

Bryant, R., & Bailey, S. (1997). *Third world political ecology*. London: Routledge.

Bucy, E. P., & Bradley, S. D. (2004). Presidential expressions and viewer emotion: Counterempathic responses to televised leader displays. *Social Science Information*, *43*(1), 59–65.

Bucy, E. P., & Bradley, S. D. (2010). What the body can tell us about politics: The use of psychophysiological measures in political communication research. In E. P. Bucy & R. L. Holbert (Eds.), *Sourcebook for political communication research: Methods, measures, and analytical techniques* (pp. 525–540). New York, NY: Taylor & Francis.

Bucy, E. P., & Grabe, M. E. (2007). Taking television seriously: A sound and image bite analysis of presidential campaign coverage, 1992–2004. *Journal of Communication*, *57*(4), 652–675.

Bucy, E. P. (2009). Biopolitics and the road ahead. *Politics and the Life Sciences*, *28*(2), 94–95.

Bucy, E. P. (2012). Biopolitics – Still descriptive? *Politics and the Life Sciences*, *31*(1–2), 1.

Cacioppo, J. T., & Decety, J. (2011). Social neuroscience: Challenges and opportunities in the study of complex behavior. *Annals of the New York Academy of Sciences*, *1224*, 162–173.

Cairns, J. Jr. (2001). Speculative scenarios about sustainable use of the planet. *The Social Contract, XL*(2), 146–152.

Caldwell, L. K. (1964). Biopolitics: Science, ethics, and public policy. *Yale Review, 56*, 1–16.

Caldwell, L. K. (1972). *In defense of earth*. Bloomington, IN: Indiana University Press.

Caldwell, L. K. (1987). *Biocracy: Public policy and the life sciences*. Boulder, CO: Westview Press.

Caldwell, L. K. (1999). Is humanity destined to self-destruct. *Politics and the Life Sciences, 18*(1), 3–14.

Camerer, C., Loewenstein, G., & Prelec, D. (2005). Neuroeconomics: How neuroscience can inform economics. *Journal of Economic Perspectives, 43*, 9–64.

Campbell, R. J., & Bickhard, M. H. (2011). Physicalism, emergence and downward causation. *Axiomathes, 21*, 33–56.

Campbell, A., Converse, P. E., Miller, W. E., & Stokes, D. E. (1960). *The American voter*. New York, NY: Wiley.

Capra, M. C., & Rubin, P. H. (2012). Rationality and utility: Economics and evolutionary psychology. In G. Saad (Ed.), *Evolutionary psychology in the business sciences* (pp. 319–338). Berlin: Springer.

Carter, A. (1999). *A radical green political theory*. London: Routledge.

Caton, H. (1990). *The politics of progress: The origins and development of the commercial republic* (pp. 1600–1835). Gainesville, FL: University of Florida Press.

Centers for Disease Control and Prevention. (2014). *Emergency preparedness and response*. Retrieved from http://www.bt.cdc.gov/preparedness/quarantine/qa.asp

Cesarini, D., Dawes, C. T., Fowler, J. H., Johannesson, M., Lichtenstein, P., & Wallace, B. (2009). Genetic variation in preferences for giving and risk taking. *The Quarterly Journal of Economics, 124*(2), 809–842.

Changeux, J. P. (1997). *Neuronal man: The biology of the mind*. Princeton, NJ: Princeton University Press.

Chapman, H. A., & Anderson, A. K. (2013). Things rank and gross in nature: A review and synthesis of moral disgust. *Psychological Bulletin, 139*(2), 300–327.

Charney, E., & English, W. (2012). Candidate genes and political behavior. *American Political Science Review, 106*(1), 1–34. doi:10.1017/S0003055411000554. Accessed on March 20, 2014.

Charney, E., & English, W. (2013). Genopolitics and the science of genetics. *American Political Science Review, 107*(2), 382–395.

Charney, E. (2008a). Genes and ideologies. *Perspectives on Politics, 6*, 299–319.

Charney, E. (2008b). Politics, genetics, and "greedy reductionism." *Perspectives on Politics, 6*(2), 337–343.

Charney, E. (2011). Political science and behavior genetics: Rethinking foundational assumptions. In S. A. Peterson & A. Somit (Eds.), *Biology and politics: The cutting edge* (Vol. 9, pp. 115–138). Research in Biopolitics. Bingley, UK: Emerald Group Publishing Limited.

Cheon, B. K., Im, D. M., Harada, T., Kim, J. S., Mathur, V. A., Scimeca, J. M. … Chiao, J. Y. (2011). Cultural influences on neural basis of intergroup empathy. *NeuroImage, 57*(2), 642–650.

Chiao, J. Y., & Ambady, N. (2007). Cultural neuroscience: Parsing universality and diversity across levels of analysis. In S. Kitayama & D. Cohen (Eds.), *Handbook of cultural psychology* (pp. 237–254). New York, NY: Guilford Press.

Chiao, J. Y., Iidaka, T., Gordon, H. L., Nogawa, J., Bar, M., Aminoff, E., & Ambay, N. (2008). Cultural specificity in amygdala response to fear faces. *Journal of Cognitive Neuroscience, 20*, 2167–2174.

Chilcote, R. (1994). *Theories of comparative politics: The search for a paradigm reconsidered* (2nd ed.). Boulder, CO: Westview Press.

Christakis, N. A., & Fowler, J. H. (2009). *Connected: The surprising power of our social networks and how they shape our lives.* New York, NY: Hachette Digital, Inc.

CIDRAP. (2007). *Odd twist in TB alert: Patient is TB researcher's son-in-law.* CIDRAP. Retrieved from http://www.cidrap.umn.edu/news-perspective/2007/05/odd-twist-tb-alert-patient-tb-researchers-son-law. Accessed on May 31.

Climate Connections. (2012). Ohio: Fracking brine gas-well waste full of radium. *Climate Connections.* Retrieved from http://climate-connections.org/2012/09/04/ohio-fracking-brine-gas-well-waste-full-of-radium/. Accessed on March 9.

Cohen, J. (1995). *How many people can the earth support?* New York, NY: Norton.

Committee on Facilitating Interdisciplinary Research. (2005). *Facilitating interdisciplinary research.* Washington, DC: National Academy Press.

Connolly, W. (1988). *Political theory and modernity.* Oxford: Basil Blackwell Ltd.

Coplan, M. J., & Masters, R. D. (2000). *Should silicofluoride be used to fluoridate municipal water?* Research Manuscript (unpublished).

Corning, P. A., & Hines, S. M., Jr. (1988). Political development and political evolution. *Politics and the Life Sciences, 6*(2), 140–172.

Corning, P. A. (1971). The biological bases of behavior and some implications for political science. *World Politics, 23*, 321–370.

Corning, P. A. (1977). Human nature redivivus. In J. R. Chapman (Ed.), *Human nature in politics, NOMOS XVII* (pp. 19–68). New York, NY: New York University Press.

Corning, P. A. (1983). *The synergism hypothesis: A theory of progressive evolution.* New York, NY: McGraw-Hill.

Corning, P. A. (1995). Synergy and self-organization in the evolution of complex systems. *Systems Research, 12*(2), 89–121.

Corning, P. A. (1996). Synergy, cybernetics and the evolution of politics. *International Political Science Review, 17*(1), 91–119.

Corning, P. (1997). Synergistic selection and the evolutionary process. *Journal of Social and Evolutionary Systems, 20*(4), 363–400.

Corning, P. A. (1998). "The synergism hypothesis": On the concept of synergy and its role in the evolution of complex systems. *Journal of Social and Evolutionary Systems, 21*(2), 133–172.

Corning, P. A. (2000). The synergism hypothesis: On the concept of synergy and its role in the evolution of complex systems. *Journal of Social and Evolutionary Systems, 24*(2), 133–172.

Corning, P. A. (2003). *Nature's magic: Synergy in evolution and the fate of humankind.* Cambridge, UK: Cambridge University Press.

Corning, P. A. (2004). The evolution of politics. In F. Wuketits & C. Antweiler (Eds.), *Handbook of evolution* (Vol. 1, pp. 1–58). Weinheim, Germany: Wiley-VCH.

Corning, P. A. (2005). *Holistic Darwinism: Synergy, cybernetics, and the bioeconomics of evolution.* Chicago, IL: University of Chicago Press.

Corning, P. A. (2008). Holistic Darwinism: The new evolutionary paradigm and some implications for political science. *Politics and the Life Sciences, 27*(1), 22–54.

Corning, P. A. (2010). *Holistic Darwinism: Synergy, cybernetics, and the bioeconomics of evolution*. Chicago, IL: University of Chicago Press.
Corning, P. A. (2011a). *The fair society and the pursuit of human justice*. Chicago, IL: University of Chicago Press.
Corning, P. A. (2011b). The synergism hypothesis: Thirty years later. *Politics and the Life Sciences*, *30*(1), 61–64.
Cory, G. A., Jr. (1999). *The reciprocal modular brain in economics and politics*. New York, NY: Plenum Press.
Costa, P. T., Jr., & McCrae, R. R. (2008). The revised NEO personality inventory (NEO-PI-R). *The SAGE Handbook of Personality Theory and Assessment*, *2*, 179–198.
Costantini, E., & Craik, K. H. (1980). Personality and politicians: California party leaders, 1960–1976. *Journal of Personality and Social Psychology*, *38*(4), 641–661.
Costanza, R. (2014). Time to leave GDP behind. *Nature*, *505*, 203–285.
Couppis, M. H., Kennedy, C. H., & Stanwood, G. D. (2008). Differences in aggressive behavior and in the mesocorticolimbic DA system between A/J and BALB/cJ mice. *Synapse*, *62*(10), 715–724.
Craig, I. W., & Halton, K. E. (2009). Genetics of human aggressive behaviour. *Human Genetics*, *126*(1), 101–113.
Cranmer, S. J., & Dawes, C. T. (2012). The heritability of foreign policy preferences. *Twin Research and Human Genetics*, *15*(1), 52–59.
Crews, D. (2008). Epigenetics and its implications for behavioral neuroendocrinology. *Frontiers in Neuroendocrinology*, *29*(3), 344–357.
Crook, J. H. (1976). Ethology and biopolitics: A discussant's view. In A. Somit (Ed.), *Biology and politics: Recent explorations* (pp. 269–276). The Hague: Mouton.
Cropsey, J. (1960). On the relation of political science and economics. *American Political Science Review*, *54*(1), 3–14.
Custers, R., & Aarts, H. (2010). The unconscious will: How the pursuit of goals operates outside of conscious awareness. *Science*, *329*(5987), 47–50.
Damasio, A. (2008). *Descartes' error: Emotion, reason and the human brain*. New York, NY: Random House.
Darwin, C. (1874). *The descent of man and selection in relation to sex*. New York, NY: A. L. Burt.
Davidson, R. J., & Begley, S. (2012). *The emotional life of your brain: How its unique patterns affect the way you think, feel, and live – and how you can change them*. New York, NY: Hudson Street Press.
Davidson, R. J., Putnam, K. M., & Larson, C. L. (2000). Dysfunction in the neural circuitry of emotion regulation: A possible prelude to violence. *Science*, *289*(5479), 591–594.
Davies, J. C. (1963). *Human nature in politics: The dynamics of political behavior*. New York, NY: Wiley.
Dawes, C., & Fowler, J. H. (2009). Partisanship, voting, and the dopamine D2 receptor gene. *The Journal of Politics*, *71*(3), 1157–1171.
Dawes, C. T., Loewen, P. J., Schreiber, D., Simmons, A. N., Flagan, T., McElreath, R., … Paulus, M. P. (2012). Neural basis of egalitarian behavior. *Proceedings of the National Academy of Sciences*, *109*(17), 6479–6483.
Dawes, R. M., Orbell, J. M., Simmons, R. T., & Van De Kragt, A. J. C. (1986). Organizing groups for collective action. *American Political Science Review*, *80*(4), 1171–1185.
Dawkins, R. (1986). *The blind watchmaker*. New York, NY: W. W. Norton & Company.
Dawkins, R. (2006). *The selfish gene*. Oxford, UK: Oxford University Press.

Dawkins, R. (2012). The descent of Edward O. Wilson. *Prospect*, May 24.
Decety, J. (2010). The neurodevelopment of empathy in humans. *Developmental Neuroscience*, *32*(4), 257–268.
De Dreu, C. K. W., Greer, L. L., Handgraaf, M. J. J., Shalvi, S., Van Kleef, G. A. et al. (2010). The neuropeptide oxytocin regulates parochial altruism in intergroup conflict among humans. *Science*, *328*(5984), 1408–1411.
Dehaene, S. (2005). Evolution of human cortical circuits for reading and arithmetic: The neuronal recycling hypothesis. In S. Dehaene, J. R. Duhamel, M. D. Hauser, & G. Rizzolatti (Eds.), *From monkey brain to human brain* (pp. 137–157). Cambridge, MA: MIT Press.
Díaz, J. L. (2010). The psychobiology of aggression and violence: Bioethical implications. *International Social Science Journal*, *61*, 233–245.
de Waal, F. B. D. (1996). *Good natured: The origins of right and wrong in humans and other animals*. Cambridge, MA: Harvard University Press.
de Waal, F. B. D. (2005). *Our inner ape*. New York, NY: Richmond Books.
de Waal, F. (2006). *Primates and philosophers: How morality evolved*. Princeton, NJ: Princeton University Press.
de Waal, F. (2009). *The age of empathy: Nature's lessons for a kinder society*. New York, NY: Three Rivers Press.
deMesquita, B. (1981). *The war trap*. New Haven, CT: Yale University Press.
Deppe, K. D., Stoltenberg, S. F., Smith, K. B., & Hibbing, J. R. (2013). Candidate genes and voter turnout: Further evidence on the role of 5-HTTLPR. *American Political Science Review*, *107*(2), 375–381.
Derntl, B., Windischerger, C., Robinson, S., Kryspin-Exner, I., Gur, R. C., Moser, E., & Habel, U. (2009). Amygdala activity to fear and anger in healthy young males is associated with testosterone. *Psychoneuroendocrinology*, *34*(5), 687–693.
DeWall, C. N., Deckman, T., Gailliot, M. T., & Bushman, B. J. (2011). Sweetened blood cools hot tempers: Physiological self-control and aggression. *Aggressive Behavior*, *37*, 73–80.
Diamond, J. M. (1992). *The third chimpanzee: The evolution and future of the human animal*. New York, NY: Harper Collins.
Diamond, J. (1997). *Guns, germs and steel: The fates of human societies*. New York, NY: W. W. Norton.
Diamond, L. (1999). *Developing democracy: Toward consolidation*. Baltimore, MD: Johns Hopkins University Press.
Diamond, J. (2005). *Collapse: How societies choose to fail or succeed*. New York, NY: Viking.
Diamond, L. (2008). The democratic rollback: The resurgence of the predatory state. *Foreign Affairs*, *87*(2), 36–48.
Diamond, J. (2012). *The world until yesterday: What can we learn from traditional societies?* New York, NY: Viking.
Dietrich, B. J., Lasley, S., Mondak, J. J., Remmel, M. L., & Turner, J. (2012). Personality and legislative politics: The big five trait dimensions among U.S. state legislators. *Political Psychology*, *33*(2), 195–210. doi:10.1111/j.1467-9221.2012.00870.x
Dietz, F., & van der Straaten, J. (1993). Economic theories and the necessary integration of ecological insights. In A. Dobson & P. Lucardie (Eds.), *The politics of nature: Explorations in green political theory* (pp. 118–144). London: Routledge.

Dmitrieva, J., Chen, C., Greenberger, E., Ogunseitan, O., & Ding, Y. C. (2011). Gender-specific expression of the DRD4 gene on adolescent delinquency, anger and thrill seeking. *Social Cognitive and Affective Neuroscience*, *6*(1), 82–89.

Dobzhansky, T., & Montagu, M. F. A. (1962). Natural selection and the mental capacities of mankind. In M. F. A. Montagu (Ed.), *Culture and the evolution of man* (pp. 148–154). New York, NY: Oxford University Press.

Dominguez Duque, J. F., Turner, R., Lewis, E. D., & Egan, G. (2010). Neuroanthropology: A humanistic science for the study of the culture–brain nexus. *Social Cognitive and Affective Neuroscience*, *5*(2–3), 138–147.

Dowdle, A., Limbocker, S., Yang, S., Sebold, K., & Stewart, P. A. (2013). *The invisible hands of political parties in presidential elections: Party activists and political aggregation from 2004 to 2012*. New York, NY: Palgrave Macmillan.

Druckman, J., Kuklinski, J., & Sigelman, L. (2009). The unmet potential of interdisciplinary research: Political psychological approaches to voting and public opinion. *Political Behavior*, *31*, 485–510.

Dryzek, J. S., & Schlosberg, D. (1995). Disciplining Darwin: Biology in the history of political science. In J. Farr, J. Dryzek, & S. Leonard (Eds.), *Political science in history: Research programs and political traditions*. Cambridge: Cambridge University Press.

Dunbar, R. I. (2012). The social brain meets neuroimaging. *Trends in Cognitive Sciences*, *16*(2), 101–102.

Dunn, J. (1990). Interpreting political responsibility: Essays 1981–1989. Oxford: Oxford University Press.

Dutton, D. (2010). *The art instinct: Beauty, pleasure, and human evolution*. Oxford: Oxford University Press.

Earley, J. (1997). *Transforming human culture: Social evolution and the planetary crisis*. Albany, NY: State University of New York Press.

Easton, D. (1965). *A systems analysis of political life*. New York, NY: Wiley.

Eaves, L. J., & Eysenck, H. J. (1974). Genetics and the development of social attitudes. *Nature*, *249*, 288–289.

Eaves, L. J., & Hatemi, P. K. (2008). Transmission of attitudes toward abortion and gay rights: Effects of genes, social learning and mate selection. *Behavioral Genetics*, *38*, 247–256.

Eaves, L. J., Hatemi, P. K., Heath, A. C., & Martin, N. G. (2011). Modeling biological and cultural inheritance. *Man is by Nature a Political Animal: Evolution, Biology, and Politics* (pp. 101–184). Chicago, IL: University of Chicago Press.

Eaves, L. J., Hatemi, P. K., Prom-Womley, E. C., & Murrelle, L. (2008). Social and genetic influences of adolescent religious attitudes and practices. *Social Forces*, *86*(4), 1621–1646.

Ebstein, R. P. (2006). The molecular genetic architecture of human personality: Beyond self-report questionnaires. *Molecular Psychiatry*, *11*, 427–445.

Eckersley, R. (1992). *Environmentalism and political theory: Toward an ecocentric approach*. Albany, NY: State University of New York Press.

Eckersley, R. (1998). Environment rights and democracy. In R. Keil, D. Bell, L. Fawcett, & P. Penz (Eds.), *Political ecology: Global and local* (pp. 353–376). London: Routledge.

Edsall, T. (2013). Are our political beliefs encoded in our DNA? *The New York Times*. Retrieved from http://opinionator.blogs.nytimes.com/2013/10/01/are-our-political-beliefs-encoded-in-our-dna/?_php=true&_type=blogs&_r=0. Accessed on April 22, 2014.

Eibl-Eibesfeldt, I. (1979). *The biology of peace and war*. New York, NY: Viking Press.

Eibl-Eibesfeldt, I. (1989). *Human ethology*. New York, NY: Aldine de Gruyter.

Eidelman, S., Crandall, C. S., Goodman, J. A., & Blanchar, J. C. (2012). Low-effort thought promotes political conservatism. *Personality and Social Psychology Bulletin, 38*(6), 808–820.

Eisenberger, N. I., Way, B. M., Taylor, S. E., Welch, W. T., & Lieberman, M. D. (2007). Understanding genetic risk for aggression: Clues from the brain's response to social exclusion. *Biological Psychiatry, 61*, 1100–1108.

Ekman, P., & Davidson, R. J. (1994). *The nature of emotion: Fundamental questions*. New York, NY: Oxford University Press.

Ellis, L. (2003). Biosocial theorizing and criminal justice policy. In A. Somit & S. A. Peterson (Eds.), *Human nature and public policy* (pp. 87–120). New York, NY: Palgrave Macmillan.

Energy Information Administration (EIA). (2012). *Energy outlook 2012*. DOE/EIA-0383. Washington, DC: Department of Energy.

Environmental Protection Agency (EPA). (2013). *Draft inventory of U.S. greenhouse gas emissions and sinks*, 1990–2011. Retrieved from http://www.epa.gov/climatechange/Downloads/ghgemissions/US-GHG-Inventory-2011.pdf. Accessed in February.

Ermer, E., Guerin, S., Cosmides, L., Tooby, J., & Miller, M. (2006). Theory of mind broad and narrow. *Social Neuroscience, 1*, 196–219.

Eulau, H. (1963). *The behavioral persuasion in politics*. New York, NY: Random House.

Falger, V. J. (Ed.). (1989). *Sociobiology and conflict: Evolutionary perspectives on group behavior*. London: Chapman and Hall.

Falger, V. (1991). The missing link in international relations theory? *Journal of Social and Biological Structures, 14*(1), 73–77.

Falger, V. (1994). Biopolitics and the study of international relations. In A. Somit & S. A. Peterson (Eds.), *Biopolitics in the mainstream*. (Vol. 2, pp. 115–134). Research in Biopolitics. London: JAI Press.

Falger, V. (1997). Human nature in modern international relations. Part I. theoretical backgrounds. In A. Somit & S. A. Peterson (Eds.), *Recent explorations in biology and politics*. (Vol. 5, pp. 155–175). Research in Biopolitics. London: JAI Press.

Falger, V. (2001a). Evolutionary world politics enriched: The biological foundations of international relations. In W. R. Thompson (Ed.), *Evolutionary interpretations of world politics* (pp. 30–51). New York, NY: Routledge.

Falger, V. (2001b). Evolution in international relations? From social Darwinism to evolutionary in the study of international relations. In S. A. Peterson & A. Somit (Eds.), *Evolutionary approaches in the behavioral sciences: Toward a better understanding of human nature*. (Vol. 8, pp. 91–118). Research in Biopolitics. Oxford, UK: JAI.

Farah, M. J. (2014). Brain images, babies, and bathwater: Critiquing critiques of functional neuroimaging. *The Hastings Center Report, 44*(Suppl. 2), s19–s30. PMID: 24634081.

Fazekas, Z., & Littvay, L. (2012). Choosing sides: The genetics of why we go with the loudest. *Journal of Theoretical Politics, 24*, 389–408.

Fenker, D. B., Heipertz, D., Boehler, C. N., Schoenfeld, M. A., Noesselt, T., Heinze, H.-J., … Hopf, J.-M. (2010). Mandatory processing of irrelevant fearful face features in visual search. *Journal of Cognitive Neuroscience, 22*(12), 2926–2938.

Ferry, L. (1995). *The new ecological order*. Chicago, IL: University of Chicago Press.

Fincher, C. L., & Thornhill, R. (2008). Assortative sociality, limited dispersal, infectious disease and the genesis of the global pattern of religion diversity. *Proceedings of the Royal Society: Biological Sciences*, *275*(1651), 2587–2594.

Fletcher, A. L. (2013). In praise of a modest postmodernism: Comment in response to Liesen and Walsh. *Politics & Life Sciences*, *32*(1), 85–89.

Forbes, G. B. (1971). Birth order and political success. *Psychological Reports*, *29*, 1239–1242.

Fowler, J. H., & Christakis, N. A. (2008). Dynamic spread of happiness in a large social network: Longitudinal analysis over 20 years in the Framingham heart study. *British Medical Journal*, *337*, a2338.

Fowler, J. H., & Dawes, C. T. (2008). Two genes predict voter turnout. *Journal of Politics*, *70*(3), 579–594.

Fowler, J. H., & Dawes, C. T. (2013). In defense of genopolitics. *American Political Science Review*, *107*(2), 362–374.

Fowler, J. H., & Schreiber, D. (2008). Biology, politics and the emerging science of nature. *Science*, *322*(5903), 912–914.

Fowler, J. H., Baker, L. A., & Dawes, C. T. (2008). Genetic variation in political participation. *American Political Science Review*, *102*, 233–248.

Fowler, J. H., Dawes, C. T., & Christakis, N. A. (2009). Model of genetic variation in human social networks. *PNAS*, *106*(6), 1720–1724.

Fowler, J. H., Loewen, P. J., Settle, J., & Dawes, C. T. (2011). Genes, games, and politcal participation. In P. K. Hatemi & R. McDermott (Eds.), *Man is by nature a political animal: Evolution, biology and politics* (pp. 207–223). Chicago, IL: University of Chicago Press.

Freeman, C. N., Cummins, E. D., Tucker, C., Ostland, J., & Hock, B. J. (2011). The addictive properties of sweetener consumption using a conditioned taste aversion animal model for dependency. *Journal of Behavioral and Neuroscience Research*, *9*(1), 1–6.

Freese, J. (2011). Integrating genomic data and social science. *Politics and the Life Sciences*, *30*(2), 88–92.

Frick, P. J., & White, S. F. (2008). The importance of callous–unemotional traits for developmental models of aggressive and antisocial behavior. *Journal of Child Psychology and Psychiatry*, *49*(4), 359–375.

Friedman, M. (1953). *Essays in positive economics*. Chicago, IL: University of Chicago Press.

Friend, J. M., & Thayer, B. A. (2011). Brain imaging and political behavior: A survey. In S. A. Peterson & A. Somit (Eds.), *Biology and politics: The cutting edge* (Vol. 9, pp. 231–255). Research in Biopolitics. Bingley, UK: Emerald Group Publishing Limited.

Fukuyama, F. (1999). *The great disruption: Human nature and the reconstitution of social order*. New York, NY: The Free Press.

Fukuyama, F. (2002). *Our posthuman future: Consequences of the biotechnology revolution*. New York, NY: Farrar, Straus, and Giroux.

Fukuyama, F. (2004). *State-building: Governance and world order in the 21st century*. Ithaca, NY: Cornell University Press.

Fukuyama, F. (2011). *The origins of political order: From prehuman times to the French revolution*. New York, NY: Farrar, Straus and Giroux.

Funke, O. (1985). Biopolitics and public policy: Controlling biotechnology. *PS: Political Science and Politics*, *8*(1), 69–77.

Funke, O. (2013). Biopolicy: Can it provide a new paradigm? In S. A. Peterson & A. Somit (Eds.), *The world of biology and politics: Organization and research areas* (Vol. 11, pp. 135–160). Research in Biopolitics. Bingley, UK: Emerald Group Publishing Limited.

Fuquay, J. (2013). Report questions long-term productivity of gas wells in Barnett Shale. *Post Telegram.* Retrieved from http://www.star-telegram.com/2013/02/12/4617558/report-questions-long-term-productivity.html. Accessed on December 2.

Gallego, A., & Oberski, D. (2012). Personality and political participation: The mediation hypothesis. *Political Behavior, 34*, 425–451.

Garrett, A., & Chang, K. (2008). The role of the amygdale in bipolar disorder development. *Development and Psychopathology, 20*, 1285–1296.

Gat, A. (2007). The return of authoritarian great powers. *Foreign Affairs, 86*(4), 59–70.

Gazzaniga, M. (2005). *The ethical brain: The science of our moral dilemmas.* New York, NY: HarperCollins.

Geiger, G. (1985). The concept of evolution and early state formation. *Politics and the Life Sciences, 3*, 163–171.

Geiger, G. (1988). On evolutionary origins and function of political power. *Journal of Social and Biological Structures, 11*, 235–250.

Gelling, C. (2013). Gut-brain link tied to overeating. *Science News, 184*(7), 15. (Referencing a study led by Ivan de Araujo et al., published in Science, August 16, 2013).

Gerber, A. S., Huber, G. A., Doherty, D., Dowling, C. M., & Ha, S. E. (2010). Personality and political attitudes: Relationships across issue domains and political contexts. *American Political Science Review, 104*, 111–133.

Gerber, A. S., Huber, G. A., Doherty, D., Dowling, C. M., & Panagopoulos, C. (2013). Big five personality traits and responses to persuasive appeals: Results from voter turnout experiments. *Political Behavior, 35*(4), 687–728.

Gerber, A., Huber, G. A., Raso, C., & Ha, S. (2009). *Personality and political behavior.* Unpublished Manuscript, Yale University.

Germino, D. (1975). The contemporary relevance of the classics for political philosophy. In F. G. Polsby (Ed.), *Handbook of political science* (Vol. 1, pp. 229–281). Reading, MA: Addison-Wesley.

Ghiselin, M. T. (1992). Biology, economics and bioeconomics. In G. Radnitzky (Ed.), *Universal economics* (pp. 71–118). New York, NY: Paragon House.

Gilbert, R. E. (2011). Presidential disability and the twenty-fifth amendment: The difficulties posed by psychological illness. *Fordham Law Review, 79*(3), 843–879.

Gilbert, R. E. (2012). The impact of presidential illness on the administration of Dwight D. Eisenhower. *Politics and the Life Sciences, 31*(1–2), 16–35.

Gilman, J. B., Lerner, B. M., Kuster, W. C., & de Gouw, J. A. (2013). Source signature of volatile organic compounds from oil and natural gas operations in northeastern Colorado. *Environmental Science & Technology, 47*(3), 1297–1305.

Glad, B. (1988). Personality, role strains and alcoholism. *Politics and the Life Sciences, 7*(1), 18–32.

Glenn, A. L., & Raine, A. (2009). Psychopathy and instrumental aggression: Evolutionary, neurobiological, and legal perspectives. *International Journal of Law and Psychiatry, 32*, 253–258.

Goetze, D. (2011). Identity challenges facing the association for politics and the life sciences. *Politics and the Life Sciences, 30*(1), 77–79.

Gold, J. I., & Shadlen, M. N. (2007). The neural basis of decision making. *Annual Review of Neuroscience*, *30*, 535–574.

Goldberg, L. R. (1990). An alternative description of personality: The big-five factor structure. *Journal of Personality and Social Psychology*, *59*(6), 1216–1229.

Gosling, S. D., Ko, S. J., Mannarelli, T., & Morris, M. E. (2002). A room with a cue: Personality judgments based on offices and bedrooms. *Journal of Personality and Social Psychology*, *82*(3), 379.

Government Accountability Office (GAO) Report. (2012). *Oil and gas information on shale resources, development, and environmental and public health risks*. GAO 12-732. Washington, DC: GAO.

Grabe, M. E., & Bucy, E. P. (2009). *Image bite politics: News and the visual framing of elections*. New York, NY: Oxford University Press.

Gray, J. A. (1987). *The psychology of fear and stress*. Cambridge, MA: Cambridge University Press.

Greenstein, F. (1975). *Personality and politics: Problems of evidence, inference, and conceptualization*. New York, NY: Norton.

Gruszczynski, M. W., Balzer, A., Jacobs, C. M., Smith, K. B., & Hibbing, J. R. (2013). The physiology of political participation. *Political Behavior*, *35*, 135–152.

Gupta, R., Koscik, T. R., Bechara, A., & Tranel, D. (2011). The amygdala and decision-making. *Neuropsychologia*, *49*, 760–766.

Gutchess, A. H., Hedden, T., Ketay, S., Aron, A., & Gabrieli., J. D. E. (2010). Neural differences in the processing of semantic relationships across cultures. *SCAN*, *5*, 254–263.

Haidt, J. (2012). *The righteous mind: Why good people are divided by politics and religion*. New York, NY: Random House.

Hajer, M. (1995). *The politics of environmental discourse: Ecological modernization and the policy process*. Oxford: Oxford University Press.

Hall, L. K. (2009). Impassioned politics. *Politics and the Life Sciences*, *28*(2), 84–89.

Hamilton, W. O. (1964). The genetic evolution of social behavior I and II. *Journal of Theoretical Biology*, *7*, 1–52.

Hammer, R., & VanBriesen, J. (2012). In fracking's wake: New rules are needed to protect our health and environment from contaminated wastewater. *NRDC Issue Brief*. Retrieved from http://www.nrdc.org/energy/files/Fracking-Wastewater-IssueBrief.pdf. Accessed in May.

Hammock, E. A. D. (2011a). Biologically constrained behavioral genetics research. *Politics and the Life Sciences*, *30*(2), 93–97.

Hammock, E. A. D. (2011b). *NSF Workshop on genes, cognition, and social behavior* (pp. 115–124). Washington, DC: National Science Foundation's Political Science Program (SBE-1037831).

Hammond, R. A., & Axelrod, R. (2006). The evolution of ethnocentrism. *The Journal of Conflict Resolution*, *50*(6), 926–936.

Han, S., & Northoff, G. (2008). Culture-sensitive neural substrates of human cognition: A transcultural neuroimaging approach. *Nature Reviews Neuroscience*, *9*, 646–654.

Hannagan, R. J., & Hatemi, P. K. (2008). The threat of genes: A comment on Evan Charney's "genes and ideologies". *Perspectives in Politics*, *6*, 329–335.

Hannagan, R. J. (2011). From genes to politics: Bridging the gap between biological and social explanations of political behavior via twin studies. In S. A. Peterson & A. Somit (Eds.), *Biology and politics: The cutting edge* (Vol. 9, pp. 139–158). Research in Biopolitics. Bingley, UK: Emerald Group Publishing Limited.

Hannagan, R. J. (2013). Biopolitics via political psychology: Comment in response to Liesen and Walsh. *Politics and the Life Sciences*, *30*(1), 90–93.

Hardin, G. (1968). The tragedy of the commons. *Science*, *162*(3859), 1243–1248.

Harkavy, R. E. (2000). Defeat, national humiliation and the revenge motif in international politics. *International Politics*, *37*(3), 345–368.

Harlan, J. M., Jr. (1955). The role of oral argument. *Cornell Law Review*, *4*, 6–15.

Harris, L. T., & Fiske, S. T. (2006). Dehumanizing the lowest of the low: Neuroimaging responses to extreme out-groups. *Psychological Science*, *17*(10), 847–853.

Harris, L. T., & Fiske, S. T. (2007). Social groups that elicit disgust are differentially processed in mPFC. *Social Cognitive and Affective Neuroscience*, *2*(1), 45–51.

Harrison, P. J., & Tunbridge, E. M. (2008). Catechol-O-methyltransferase (COMT): A gene contributing to sex differences in brain function, and to sexual dimorphism in the predisposition to psychiatric disorders. *Neuropsychopharmacology*, *33*, 3037–3045.

Hartigan, R. S. (1995). Natural law, nature, and biology: A modern synthesis. In A. Somit & J. Losco (Eds.), *Research in biopolitics: Human nature and politics* (Vol. 3). Greenwich, CT: JAI Press.

Hatemi, P. K., & McDermott, R. (2010). Emerging models of collaboration in political science: Changes, benefits, and challenges. *PS: Political Science and Politics*, *43*(1), 49–58.

Hatemi, P. K., & McDermott, R. (2011). *Man is by nature a political animal: Evolution, biology, and politics*. Chicago, IL: University of Chicago Press.

Hatemi, P. K., & McDermott, R. (2012a). The political psychology of biology, genetics, and behavior. *Political Psychology*, *33*, 307–312.

Hatemi, P. K., & McDermott, R. (2012b). The genetics of politics: Discovery, challenges, and progress. *Trends in Genetics*, *28*(10), 525–533.

Hatemi, P. K. (2011). NSF Workshop on genes, cognition, and social behavior (pp. 125–137). Washington, DC: National Science Foundation's Political Science Program SBE-1037831.

Hatemi, P. K., Alford, J. R., Hibbing, J. R., Martin, N. G., & Eaves, L. J. (2009). Is there a "party" in your genes? *Political Research Quarterly*, *62*(3), 584–600.

Hatemi, P. K., McDermott, R., Eaves, L. J., Kendler, K. S., & Neale, M. C. (2013). Fear as a disposition and an emotional state: A genetic and environmental approach to out-group political preferences. *American Journal of Political Science*, *57*, 279–293.

Hatemi, P. K., Funk, C. L., Medland, S. E., Maes, H. M., Silberg, J. L., Martin, N. G., & Eaves, N. J. (2009). Genetic and environmental transmission of political attitudes over a life time. *The Journal of Politics*, *71*, 1141–1156.

Hatemi, P. K., Alford, J., Hibbing, J., Keller, M., Martin, N., Medland, S., & Eaves, L. J. (2010). Not by twins alone: Using the extended twin family designed to investigate the genetic basis of political beliefs. *American Journal of Political Science*, *54*(3), 798–814.

Hatemi, P. K., Gillespie, N. A., Eaves, L. J., Maher, B. S., Webb, B. T., Heath, A. C., … Martin, N. G. (2011). Genome-wide analysis of liberal and conservative political attitudes. *Journal of Politics*, *73*(1), 1–15.

Hatemi, P. K., Medland, S. E., Klemmensen, R., Oskarsson, S., Littvay, L., Dawes, C. T., … Martin, N. G. (2014). Genetic influences on political ideologies: Twin analyses of 19 measures of political ideologies from five democracies and genome-wide findings from three populations. *Behavior Genetics*. doi:10.1007/s10519-014-9648-8.

Hatemi, P. K., McDermott, R., Eaves, L. J., Kendler, K. S., & Neale, M. C. (2013). Fear as a disposition and an emotional state: A genetic and environmental approach to out-group political preferences. *American Journal of Political Science*, *57*(2), 279–293.

Hatemi, P. K., Medland, S. E., Morley, K. I., Heath, A. C., & Martin, N. G. (2007). The genetics of voting: An Australian twin study. *Behavior Genetics*, *37*, 435–448.

Hayward, T. (1998). *Political theory and ecological values*. New York, NY: St. Martin's Press.

Hehman, E., Carpinella, C. M., Johnson, K. L., Leitner, J. B., & Freeman, J. B. (2014). Early processing of gendered facial cues predicts the electoral success of female politicians. Social Psychological and Personality Science. doi:10.1177/1948550614534701.

Hertsgaard, M. (1999). *Earth odyssey: Around the world in search of our environmental future*. New York, NY: Broadway Books.

Hetherington, M. J., & Weiler, J. D. (2009). *Authoritarianism and polarization in American politics*. New York, NY: Cambridge University Press.

Hibbing, M. V., Ritchie, M., & Anderson, M. R. (2011). Personality and political discussion. *Political Behavior*, *33*, 601–624.

Hibbing, J. R., Smith, K. B., & Alford, J. R. (2013). *Predisposed: liberals, conservatives, and the biology of political differences*. London: Routledge.

Hill, R. A., & Dunbar, R. I. (2003). Social network size in humans. *Human Nature*, *14*(1), 53–72.

Hines, S. (1978). Global interdependence and value diffusion. In S. K. Sharma (Ed.), *Aspects of development* (pp. 213–241). New Delhi: Concept Publishers.

Hines, S. M. (1979). Evolutionary epistemology and political knowledge. In M. Falco (Ed.), *Through the looking glass* (pp. 329–365). Washington, DC: University Press of America.

Hines, S. M. (1982a). Is synthesis philosophically possible? The paradigm problem in the philosophy of social science. In W. T. Bluhm (Ed.), *Political science and philosophy of science* (pp. 25–63). Durham, NC: Carolina Academic Press.

Hines, S. M. (1982b). Biopolitics and the evolution of inquiry in political science. *Politics and the Life Sciences*, *1*(1), 5–16.

Hines, S. M. (1982). Is synthesis philosophically possible? The paradigm problem in the philosophy of social science. In W. T. Bluhm (Ed.), *Political science and philosophy of science* (pp. 25–63). Durham, NC: Carolina Academic Press.

Hines, S. (1983). The origins of the state. In H. Flohr & W. Tommesemann (Eds.), *Politak und Biologie* (pp. 68–79). Berlin: Verlag Paul Parey.

Hines, S. M. (1998). Harold Lasswell: Biopolitics pioneer (Unpublished manuscript).

Hodgson, G. M. (1993). *Economics and evolution: Bringing life back into economics*. Cambridge: Polity Press.

Hodson, G., & Costello, K. (2007). Interpersonal disgust, ideological orientations, and dehumanization as predictors of intergroup attitudes. *Psychological Science*, *18*(8), 691.

Homer-Dixon, T. (1999). *Environment, scarcity, and violence*. Princeton, NJ: Princeton University Press.

Hopple, G. (1980). *Political psychology and biopolitics: Assessing and predicting elite behavior in foreign policy crises*. Boulder, CO: Westview.

Hoptman, M. J., & Antonius, D. (2011). Neuroimaging correlates of aggression in schizophrenia: An update. *Current Opinion in Psychiatry*, *24*(2), 100–106.

Howarth, R. W., Santoro, R., & Ingraffea, A. (2011). Methane and the greenhouse-gas footprint of natural gas from shale formations. *Climatic Change*, *106*(4), 679–690.

Hsu, M., Bhatt, M., Adolphs, R., Tranel, D., & Camerer, C. F. (2005). Neural system responding to degrees of uncertainty in human decision-making. *Science*, *310*, 1681–1683.

Hughes, D. (2011). *Life cycle greenhouse gas emissions from shale gas compared to coal: An analysis of two conflicting studies*. Postcarbon Institute. Retrieved from http://www.postcarbon.org/reports/PCI-Hughes-NETL-Cornell-Comparison.pdf. Accessed in July.

Hughes, D. (2012, May). *Will natural gas fuel America in the 21st century?* Postcarbon Institute. Retrieved from http://www.postcarbon.org/report/331901-report-will-natural-gas-fuel-america

Huntington, S. (1968). *Political order in changing societies*. New Haven, CT: Yale University Press. Reissued in 2006 with a Forward by Francis Fukuyama.

Häyry, M. (2010). Neuroethical theories. *Cambridge Quarterly of Healthcare Ethics*, *19*, 165–178.

Ikenberry, G. J. (2014). The illusion of geopolitics: The enduring power of the liberal order. *Foreign Affairs*, *93*(3), 80–91.

Inbar, Y., Pizarro, D., Iyer, R., & Haidt, J. (2012). Disgust sensitivity, political conservatism, and voting. *Social Psychological and Personality Science*, *3*(5), 537–544.

Ingram, H., Schneider, A. L., & DeLeon, P. (2007). Social construction and policy design. In P. A. Sabatier (Ed.), *Theories of the policy process* (pp. 93–128). Boulder, CO: Westview Press.

Ivers, D. J., & Balash, D. (1999). Women, men, and leadership: An evolutionary biological explanation for differing strategies for obtaining and using power positions in politics. Paper presented at annual meeting of the Association for Politics and the Life Sciences, Atlanta, September.

Jasny, B. R., Kelner, K. L., & Pennisi, E. (2008). From genes to social behavior. *Science*, *322*(5903), 891.

Jennings, M. K., & Markus, G. B. (1984). Partisan orientations over the long haul: Results from the three wave political socialization panel study. *American Political Science Review*, *78*(4), 1000–1018.

Jennings, M. K., & Niemi, R. G. (1981). *Generations and politics: A panel study of young adults and their parents*. Princeton, NJ: Princeton University Press.

Johnson, G. R. (1986). Kin selection, specialization, and patriotism: An integrating theory. *Politics and the Life Sciences*, *4*(127–140), 149–154.

Johnson, G. R. (1987). In the name of the Fatherland. *International Political Science Review*, *8*, 165–174.

Johnson, G. R. (1995). The evolutionary origins of government and politics. In A. Somit & J. Losco (Eds.), *Human nature and politics*. (Vol. 3, pp. 243–305). Research in Biopolitics. London: JAI Press.

Johnson, G. R. (2011). Politics and the life sciences. *Politics and the Life Sciences*, *30*(2), 43–64.

Johnson, T. R., Black, R. C., & Wedeking, J. (2009). Pardon the interruption: An empirical analysis of Supreme Court justices' behavior during oral arguments. *Loyola Law Review*, *55*, 331.

Johnson, T. R., Black, R. C., Goldman, J., & Treul, S. A. (2009). Inquiring minds want to know: Do justices tip their hands with questions at oral argument in the US Supreme Court. *Washington University Journal of Law and Policy*, *29*, 241.

Jonas, H. (1984). *The imperative of responsibility: In search of an ethics for the technological age*. Chicago, IL: University of Chicago Press.

Joseph, J. (2010). The genetics of political attitudes and behavior: Claims and refutations. *Ethical Human Psychology and Psychiatry*, *12*(3), 200–217.

Jost, J. T. (2006). The end of the end of ideology. *American Psychologist*, *61*, 651–670.

Jost, J. T., Glaser, J., Kruglanski, A. W., & Sulloway, F. J. (2003). Political conservatism as motivated social cognition. *Psychological Bulletin*, *129*, 339–375.

Judge, T. A., & Cable, D. M. (2004). The effect of physical height on workplace success and income: Preliminary test of a theoretical model. *Journal of Applied Psychology*, *89*(3), 428–436.

Kagan, R. (2012). *The world America made*. New York, NY: Alfred Knopf.

Kahler, M. (2002). The state of the state in world politics. In I. Katznelson & H. V. Milner (Eds.), *Political science: State of the discipline* (pp. 56–83). New York, NY: W. W. Norton.

Kalmoe, N. P. (2013). From fistfights to firefights: Trait aggression and support for state violence. *Political Behavior*, *35*, 311–330.

Kam, C. D., & Simas, E. N. (2010). Risk orientations and policy frames. *The Journal of Politics*, *72*, 381–396.

Kam, C. D. (2012). Risk attitudes and political participation. *American Journal of Political Science*, *56*(4), 817–836.

Kanai, R., & Rees, G. (2011). The structural basis of inter-individual differences in human behaviour and cognition. *Nature Reviews Neuroscience*. doi:10.1038/nrn3000. pp. 231–242.

Kanai, R., Feilden, T., Firth, C., & Rees, G. (2011). Political orientations are correlated with brain structure in young adults. *Current Biology*, *21*, 677–680.

Kaplan, R. (1996). *The ends of the earth: A journey at the dawn of the 21st century*. New York, NY: Random House.

Kaplan, R. (2012). *The revenge of geography: What the map tells us about coming conflicts and the battle against fate*. New York, NY: Random House.

Kaplan, J. T., Freedman, J., & Iacoboni, M. (2007). Us versus them: Political attitudes and party affiliation influence neural response to faces of presidential candidates. *Neuropsychologia*, *45*, 55–64.

Kasper, C., & Voelkl, B. (2009). A social network analysis of primate groups. *Primates*, *50*(4), 343–356.

Kassiola, J. (1990). *The death of industrial civilization: The limits to economic growth and the repoliticization of advanced industrial society*. Albany, NY: State University of New York Press.

Kato, J., Ide, H., Kabashima, I., Kadota, H., Takano, K., & Kansaku, K. (2009). Neural correlates of attitude change following positive and negative advertisements. *Frontiers of Behavioral Neuroscience*, *3*, 6–12.

Kelsey, T. W., Metcalf, A., & Salcedo, R. (2012). Marcellus shale: Land ownership, local voice, and the distribution of lease and royalty dollars. CECD Research Paper Series, Pennsylvania State University.

Kidd, D. C., & Castano, E. (2013). Reading literary fiction improves theory of mind. *Science*, *342*(6156), 377–380.

Kieffer, G. (1977). *Biology and human affairs*. Champaign, IL: Stipes Publishing Company.

Kim, J. (2006). Emergence: Core ideas and issues. *Synthese*, *151*(3), 547–559.

Kim, M. J., Loucks, R. A., Neta, M., Davis, F. C., Oler, J. A., Mazzulla, E. C., & Whelen, P. J. (2010). Behind the mask: The influence of mask-type on amygdala response to fearful faces. *Social Cognitive and Affective Neuroscience*, *5*(4), 363–368.

Kimmel, C. M., Stewart, P. A., & Schreckhise, W. D. (2012). Of closed minds and open mouths: Indicators of supreme court justice votes during the 2009 and 2010 sessions. *The Forum*, *10*, 2.

Kimonis, E. R., Frick, P. J., Skeem, J. L., Marsee, M. A., Cruise, K., Munoz, L. C., … Morris, A. S. (2008). Assessing callous–unemotional traits in adolescent offenders: Validation of the inventory of callous–unemotional traits. *International Journal of Law and Psychiatry*, *31*, 241–252.

Kitayama, S., & Park, J. (2010). Cultural neuroscience of the self: Understanding the social grounding of the brain. *Social Cognitive and Affective Neuroscience*, *5*(2–3), 111–129.

Kitchen, W., & Peterson, S. A. (1996). Hemispheric dominance and elite political behavior. In A. Somit & S. Peterson (Eds.), *Research in biopolitics* (Vol. 4, pp. 139–150). Greenwich, CT: JAI Press.

Kitchen, W. (1986). The dialectic of dereliction: A neuro-linguistic analysis of the language of the watergate group. In E. White & J. Losco (Eds.), *Biology and bureaucracy*. Lanham, MD: University Press of America.

Klemmensen, R., Hatemi, P. K., Hobolt, S. B., Petersen, I., Skytthe, A., & Nørgaard, A. S. (2012). The genetics of political participation, civic duty, and political efficacy across cultures: Denmark and the United States. *Journal of Theoretical Politics*, *24*(3), 409–427.

Kolbert, E. (2014). *The sixth extinction: An unnatural history*. New York, NY: Henry Holt & Co.

Koppensteiner, M., & Grammer, K. (2010). Motion patterns in political speech and their influence on personality ratings. *Journal of Research in Personality*, *44*, 374–379.

Koppensteiner, M., & Grammer, K. (2011). Body movements of male and female speakers and their influence on perceptions of personality. *Personality and Individual Differences*, *51*(6), 743–747.

Koppensteiner, M. (2011). Perceiving personality in simple motion cues. *Journal of Research in Personality*, *45*(4), 358–363.

Koppensteiner, M. (2013). Motion cues that make an impression: Predicting perceived personality by minimal motion information. *Journal of Experimental Social Psychology*, *49*(6), 1137–1143.

Koscik, T. R., & Tranel, D. (2011). The human amygdala is necessary for developing and expressing normal interpersonal trust. *Neuropsychologia*, *49*, 602–611.

Kosfeld, M. (2007). Trust in the brain: Neurobiological determinants of human social behaviour. *EMBO Reports*, *8*, s44–s47.

Kosfeld, M., Heinrichs, M., Zak, P. J., Fischbacher, U., & Fehr, E. (2005). Oxytocin increases trust in humans. *Nature*, *435*(7042), 673–676.

Kress, P. (1983). Political theorizing in the late twentieth century: Foci, loci, and agendas. In J. Nelson (Ed.), *What should political theory be now?* (pp. 103–126). Albany, NY: State University of New York Press.

Kuhn, T. S. (1996). *The structure of scientific revolutions*. Chicago, IL: University of Chicago Press.

Laitin, D. D. (2002). Comparative politics: The state of the subdiscipline. In I. Katznelson & H. V. Milner (Eds.), *Political science: State of the discipline* (pp. 630–659). New York, NY: W. W. Norton.

Lamm, R., & Blank, R. H. (2007). *Condition critical: A new moral vision for American health care*. Denver, CO: Fulcrum Press.

Lamm, C., Meltzoff, A. N., & Decety, J. (2010). How do we empathize with someone who is not like us? A functional magnetic resonance imaging study. *Journal of Cognitive Neuroscience*, *22*(2), 362–376.

LaPonce, J. A. (1976). The left-hander and politics. In A. Somit (Ed.), *Biology and politics*. Paris: Mouton.

Lasswell, H. (1968). The future of the comparative method. *Comparative Politics*, *1*(1), 3–18.

Lasswell, H. (1970). *The analysis of political behavior: An empirical approach*. Hamden, CT: Archon Books.

Laszlo, E. (Ed.). (1991). *The new evolutionary paradigm* (Vol. 2). The World Futures General Evolution Studies. New York, NY: Gordon and Breach Science Publishers.

LaVaque-Manty, M. (2006). *Nature's new constraints? Political theory and the life sciences boom*. Retrieved from http://www-personal.umich.edu/~mmanty/research/Natures_constraints.pdf

Lawford-Smith, H. (2012). Peter Corning: The fair society: The science of human nature and the pursuit of social justice. *Biology and Philosophy*, 27, 313–320.

Lehner, P. (1996). *Handbook of ethology methods*. Cambridge: Cambridge University Press.

Leiss, W. (1976). The limits to satisfaction: An essay on the problem of needs and commodities. Toronto, ON: University of Toronto Press.

Lemke, T. (2010). From state biology to the government of life: Historical dimensions and contemporary perspectives of "biopolitics". *Journal of Classical Sociology*, *10*, 421–438.

Lemke, T. (2011). *Bio-politics: An advanced introduction*. New York, NY: NYU Press.

L'Etang, H. (1988). The effect of drugs on political decisions. *Politics and the Life Sciences*, *7*(1), 12–17.

Lewontin, R. (1979). Sociobiology as an adaptationist program. *Behavioral Scientist*, *24*, 5–14.

Li, Y. J., Cohen, A. B., Weeden, J., & Kenrick, D. T. (2010). Mating competitors increase religious beliefs. *Journal of Experimental Social Psychology*, *46*(2), 428–431.

Lieberman, M., Schreiber, D., & Ochsner, K. (2003). Is political sophistication like learning to ride a bicycle? How cognitive neuroscience can inform research on political thinking. *Political Psychology*, *24*(4), 681–704.

Liesen, L. T., & Walsh, M. B. (2013). Embracing intellectual humility: A response to the biopolitics forum. *Politics & Life Sciences*, *32*(1), 99–101.

Liesen, L. T. (2012). The competing meanings of "biopolitics" in political science: Biological and postmodern approaches to politics. *Politics and the Life Sciences*, *31*(1–2), 2–15.

Little, A. C., Burriss, R. P., Jones, B. C., & Roberts, S. C. (2007). Facial appearance affects voting decisions. *Evolution and Human Behavior*, *28*(1), 18–27.

Littvay, L., Weith, P. T., & Dawes, C. T. (2011). Sense of control and voting: A genetically-driven relationship. *Social Science Quarterly*, *92*(5), 1236–1252.

Liu, J., Raine, A., Venables, P. H., & Melnick, S. A. (2004). Malnutrition at age 3 years and externalizing behavior problems at ages 8, 11 and 17 years. *American Journal of Psychiatry*, *16*, 2005–2013.

Livingston, R. W., & Pearce, N. A. (2009). The teddy-bear effect: Does having a baby face benefit black chief executive officers. *Psychological Science*, *20*(10), 1229–1236.

Lodge, M., & Somit, A. (2013). The program at the state University of New York at Stony Brook. In S. A. Peterson, & A. Somit (Eds.), *The world of biology and politics: Organization and research areas* (Vol. 11, pp. 31–33). Research in Biopolitics. Bingley, UK: Emerald Group Publishing Limited.

Lodge, M., Cross, D. V., Tursky, B., & Tanenhaus, J. (1975). The psychophysical scaling and validation of a political support scale. *American Journal of Political Science*, *19*, 611–649.

Lomberg, B. (2004). *Global crises, Global solutions*. Cambridge: Cambridge University Press.

Lomborg, B. (Ed.). (2004). *Global crises, global solutions*. Cambridge, UK: Cambridge University Press.

Long, W. J., & Brecke, P. (2003). The emotive causes of recurrent international conflicts. *Politics and the Life Sciences*, *22*(1), 24–35.

Lopez, A. C., & McDermott, R. (2012). Adaptation, hereditability and the emergence on evolutionary political science. *Political Psychology*, *33*(3), 343–362.

Losco, J., & Day Baird, D. (1982). The impact of sociobiology on political science. *American Behavioral Scientist*, *25*(3), 335–360.

Losco, J. (1981). Ultimate vs. proximate explanation: Explanatory modes in sociobiology and the social sciences. *Journal of Social and Biological Structures*, *4*(4), 329–346.

Losco, J. (2001). Evolutionary theory and political thought. In S. A. Peterson & A. Somit (Eds.), *Evolutionary approaches in the behavioral sciences: Toward a better understanding of human nature*. (Vol. 8, pp. 161–180). Research in Biopolitics. Oxford, UK: JAI.

Losco, J. (2011). From outrage to orthodoxy? *Politics and the Life Sciences*, *30*(1), 80–84.

Losin, E. A., Dapretto, M., & Iacoboni, M. (2010). Culture and neuroscience: Additive or synergistic? *Social Cognitive and Affective Neuroscience*, *5*(2–3), 148–158.

Lowi, T. J. (1972). Four systems of policy, politics, and choice. *Public Administration Review*, *32*(4), 298–310.

Lupia, A. (2011a). Genes, cognition, and social behavior. *Politics and the Life Sciences*, *30*(2), 65–87.

Lupia, A. (2011b). NSF workshop on genes, cognition, and social behavior. Washington, DC: National Science Foundation's Political Science Program SBE-1037831.

Lynn, R., & Vanhannen, T. (2002). *IQ and the wealth of nations*. Westport, CT: Praeger.

Lynn, R., & Vanhannen, T. (2006). *IQ and global inequality*. Augusta, GA: Washington Summit Publishers.

Lynn, R., & Vanhannen, T. (2012). *Intelligence: A unifying construct for the social sciences*. London: Ulster Institute for Social Research.

Ma, Q., Shen, Q., Xu, Q., Li, D., Shu, L., & Weber, B. (2011). Empathetic responses to others' gains and losses: An electrophysiological investigation. *NeuroImage*, *54*(3), 2472–2480.

Macpherson, C. B. (1962). *The political theory of possessive individualism: Hobbes to Locke*. Oxford: Clarendon Press.

Madsen, D. (1986). Power seekers are different: Further biochemical evidence. *American Political Science Review*, *80*, 261–269.

Mair, P. (1996). Comparative politics: An overview. In R. Goodin & H. D. Klingemann (Eds.), *A new handbook of political science* (pp. 309–335). Oxford: Oxford University Press.

Marcus, G. E. (2000). Emotions in politics. *Annual Review of Political Science*, *3*, 221–250.

Marcus, G. E. (2013). *Political psychology: Neuroscience, genetics, and politics*. New York, NY: Oxford University Press.

Marmot, M. (2004). *The status syndrome: How your social standing affects your health and life expectancy*. London: Bloomsbury.

Marras, A. (2006). Emergence and reduction: Reply to Kim. *Synthese*, *151*(3), 561–569.

Martin, N. G., Eaves, L. J., Heath, A. C., Jardine, R., Feingold, L. M., & Eysenck, H. J. (1986). Transmission of social attitudes. *Proceedings of the National Academy of Sciences* (15 June), 4364–4368.

Maslow, A. H. (1943). A theory of human motivation. *Psychological Review*, *50*, 370–396.

Masters, R. D., & Carlotti, S. J., Jr. (1994). Gender differences in responses to political leaders. In L. Ellis (Ed.), *Social stratification and socioeconomic inequality*. Boulder, CO: Praeger Press.

Masters, R. D., & Coplan, M. (2001). Toxins, brain chemistry, and behavior. *Optimal Center Newsletter*, p. 262. Retrieved from www.mercola.com/2001/oct/10/toxins.html. Accessed on October 10.

Masters, R. D., & Sullivan, D. G. (1993). Nonverbal behavior and leadership: Emotion and cognition in political attitudes. In S. Iyengar & W. J. McQuire (Eds.), *Explorations in political psychology*. Durham, NC: Duke University Press.

Masters, R. D., & Way, B. (1996). Experimental methods and attitudes toward leaders. In A. Somit & S. A. Peterson (Eds.), *Research in biopolitics* (Vol. 4, pp. 61–98). London: JAI Press.

Masters, R. D. (1964). World politics as a primitive political system. *World Politics*, *16*, 595–619.

Masters, R. D. (1975). Politics as a biological phenomenon. *Social Science Information*, *14*(2), 7–63.

Masters, R. D. (1978). Attention structures and political campaigns. Paper presented at the Annual Meeting of American Political Science Association, New York, September.

Masters, R. D. (1989). *The nature of politics*. New Haven, CT: Yale University Press.

Masters, R. D. (1990). Evolutionary biology and political theory. *American Political Science Review*, *84*(1), 195–210.

Masters, R. D. (1993). Why study serotonin, social behavior, and the law? In R. D. Masters & M. T. McGuire (Eds.), *The neurotransmitter revolution: Serotonin, social behavior, and the law*. Carbondale, IL: Southern Illinois University Press.

Masters, R. D. (1994). Conclusion: Primate politics and political theory. In G. Schubert & R. Masters (Eds.), *Primate politics*. Lanham, MD: University Press of America.

Masters, R. D. (2002). MacLean's evolutionary neuroethology: Environmental pollution, brain chemistry, and violent crime. In G. A. Corey, Jr., & R. Gardner, Jr. (Eds.), *The evolutionary neuroethology of Paul MacLean* (pp. 275–296). Westport, CN: Praeger.

Masters, R. D. (2003). The social implications of evolutionary psychology: Linking brain chemistry, toxins and violent crime. In R. W. Bloom & N. Dess (Eds.), *Evolutionary psychology and violence* (pp. 23–56). Westport, CN: Praeger.

Masters, R. D., & Sullivan, D. G. (1989). Nonverbal displays and political leadership in France and the United States. *Political Behavior*, *11*, 121–153.

Masters, R. D., Way, B., Hone, B. T., Grelotti, D. J., Gonzalez, D., & Jones, D. (1998). Neurotoxicology and violence. *Vermont Law Review*, *22*, 358–382.

Masters, R. D., Sullivan, D. G., Lanzetta, J. T., McHugo, G. J., & Englis, B. G. (1986). The facial displays of leaders: Toward an ethology of human politics. *Journal of Social and Biological Structures*, *9*, 319–343.

Matsumoto, D., & Willingham, B. (2009). Spontaneous facial expressions of emotion of congenitally and noncongenitally blind individuals. *Journal of Personality and Social Psychology*, *96*(1), 1–6.

Matsumoto, D., Keltner, D., Shiota, M. N., O'Sullivan, M., & Frank, M. (2010). Facial expressions of emotion. In M. Lewis, J. M. Haviland-Jones, & L. F. Barrett (Eds.), *Handbook of emotions* (3rd ed., pp. 211–234). New York, NY: The Guilford Press.

Mattes, K., Spezio, M., Kim, H., Todorov, A., Adolphs, R., & Alvarez, R. M. (2010). Predicting election outcomes from positive and negative trait assessments of candidate images. *Political Psychology*, *31*(1), 41–58.

Maynard-Smith, J., & Harper, D. (2003). *Animal signals*. New York, NY: Oxford University Press.

Mayr, E. (1960). The emergence of evolutionary novelties. In S. Tax (Ed.), *Evolution after Darwin* (Vol. 1, pp. 349–380). The Evolution of Life. Chicago, IL: University of Chicago Press.

Mazur, A., & Mueller, U. (1996). Facial dominance. In A. Somit & S. A. Peterson (Eds.), *Research in biopolitics* (Vol. 4, pp. 99–112). London: JAI Press.

Mazur, A. (2005). *Biosociology of dominance and deference*. Lanham, MD: Rowman & Littlefield.

McCrae, R. R., & Costa, P. T., Jr. (1987). Validation of the five-factor model of personality across instruments and observers. *Journal of Personality and Social Psychology*, *52*(1), 81–90.

McCrae, R. R., & Costa, P. T., Jr. (2008). The five-factor theory of personality. In R. W. O. P. John (Ed.), *Handbook of personality: Theory and research* (pp. 159–181). New York, NY: Guilford Press.

McCrae, R. R., Jang, K. L., Livesley, W. J., Riemann, R., & Angleitner, A. (2001). Sources of structure: Genetic, environmental, and artifactual influences on the covariation of personality traits. *Journal of Personality*, *69*(4), 511–535.

McDermott, R. (2004). The feeling of rationality: The meaning of neuroscience advances for political science. *Perspectives on Politics*, *2*, 691–706.

McDermott, R. (2009). Mutual interests: The case of increasing dialogue between political science and neuroscience. *Political Research Quarterly*, *62*(3), 571–583.

McDermott, R. (2011a). Combining social and biological approaches to political behaviors. *Politics and the Life Sciences*, *30*(2), 98–102.

McDermott, R. (2011b). Hormones and politics. In P. K. Hatemi & R. McDermott (Eds.), *Man is by nature a political animal: Evolution, biology, and politics* (pp. 247–260). Chicago, IL: University of Chicago Press.

McDermott, R., Johnson, D., Cowden, J., & Rosen, S. (2007). Testosterone and aggression in a simulated crisis game. *The Annals of the American Academy of Political and Social Science*, *614*, 15–33.

McDermott, R., Tingley, D., Cowden, J., Frazzetto, G., & Johnson, D. D. P. (2009). Monoamine oxidase A gene (MAOA) predicts behavioral aggression following provocation. *Proceedings of the National Academy of Sciences*, *106*(7), 2118–2123.

McHugo, G. J., Lanzetta, J. T., & Bush, L. K. (1991). The effect of attitudes on emotional reactions to expressive displays of political leaders. *Journal of Nonverbal Behavior*, *15*(1), 19–41.

McHugo, G. J., Lanzetta, J. T., Sullivan, D. G., Masters, R. D., & Englis, B. G. (1985). Emotional reactions to a political leader's expressive displays. *Journal of Personality and Social Psychology*, *49*(6), 1513–1529.

McKenzie, L., Witter, R., Newman, L., & Adgate, J. (2012). Human health risk assessment of air emissions from development of unconventional natural gas resources. *Science of the Total Environment*, *424*, 79–87.

McManus, J., Laughlin, C. D., Jr., & d'Aquili, E. G. (1979). Concepts, methods and conclusions. In E. G. d'Aquili (Ed.), *The spectrum of ritual: A biogenetic structural analysis*. New York, NY: Columbia University Press.

Mechanic, D., & Hansell, S. (1987). Adolescent competence, psychological well-being and self-assessed physical health. *Journal of Health and Social Behavior*, *28*, 364–374.

Medland, S. E., & Hatemi, P. K. (2009). Political science, biometric theory, and twin studies: A methodological introduction. *Political Analysis*, *17*, 191–214.

Mehta, P. H., & Beer, J. (2010). Neural mechanisms of the testosterone–aggression relation: The role of orbitofrontal cortex. *Journal of Cognitive Neuroscience*, *22*(10), 2357–2368.

Merriam, C. E. (1926). Progress in political research. *American Political Science Review*, *20*, 1–17.

Mertins, V., Schote, A. B., Hoffeld, W., Griessmair, M., & Meyer, J. (2011). Genetic susceptibility for individual cooperation preferences: The role of monoamine oxidase A gene (MAOA) in the voluntary provision of public goods. PloS One, *6*(6), e20959.

Meyer-Lindenberg, A., Buckholtz, J. W., Kolachana, B., Hariri, R., Pezawas, L., Blazi, G., ... Weinberger, D. R. (2006). Neural mechanisms of genetic risk for impulsivity and violence in humans. *Proceedings of the National Academy of Science*, *103*, 6269–6274.

Miller, W. (1991). Testimony before the NSF biological and behavioral science task force on reorganization. *PS: Political Science and Politics*, *24*, 7–9.

Miller, G. F. (2000). *The mating mind: How sexual choice shaped the evolution of human nature*. New York, NY: Doubleday Books.

Miller, G. F. (2007). Sexual selection for moral virtues. *Quarterly Review of Biology*, *82*(2), 97–125.

Miller, G. (2009). *Spent: Sex, evolution, and consumer behavior*. New York, NY: Viking.

Mills, I. H. (1973). Biological factors in international relations. *Yearbook of World Affairs*, *17*, 316–331.

Modelski, G. (2001). Evolutionary world politics: Problems of scope and method. In W. R. Thompson (Ed.), *Evolutionary interpretations of world politics* (pp. 16–29). New York, NY: Routledge.

Modinos, G., Ormel, J., & Aleman, A. (2010). Individual differences in dispositional mindfulness and brain activity involved in reappraisal of emotion. *Social Cognitive and Affective Neuroscience*, *5*(4), 369–377.

Mondak, J. J., & Halperin, K. D. (2008). A framework for the study of personality and political behavior. *British Journal of Political Science*, *38*, 335–362.

Mondak, J. J., Hibbing, M. V., Canache, D., Seligson, M. A., & Anderson, M. R. (2010). Personality and civic engagement: An integrative framework for the study of trait effects on political behavior. *American Political Science Review*, *104*(1), 85–110.

Morgenthau, H. (1967). *Politics among nations* (4th ed.). New York, NY: Alfred Knopf.

Morris, K. V. (2012, October 1). Lamarck and the missing Lnc: Epigenetic changes accrued over an organism's lifetime may leave a permanent heritable mark on the genome, through the help of long noncoding RNAs. *The Scientist*, *26*(10).

Moynihan, D. (1993). *Pandaemonium: Ethnicity in international politics*. Oxford: Oxford University Press.

Mueller, J. (2014). Did history end? Assessing the Fukuyama thesis. *Political Science Quarterly*, *129*(1), 35–54.

Munro, W. B. (1928). Physics and politics – An old analogy revised. *American Political Science Review*, *22*, 1–11.

Murray, G. R., & Murray, S. M. (2011). Caveman executive leadership: Evolved leadership preferences and biological sex. In G. Saad (Ed.), *Evolutionary psychology in the business sciences* (pp. 135–163). New York, NY: Springer.

Mutz, D. C., & Reeves, B. (2005). The new videomalaise: Effects of televised incivility on political trust. *American Political Science Review*, *99*(1), 1–15.

Mutz, D. C. (2007). Effects of "in-your-face" television discourse on perceptions of a legitimate opposition. *American Political Science Review*, *101*(4), 621–635.

National Institute on Drug Abuse. (2010). Impulsive-antisocial personality traits linked to a hypersensitive brain reward system. *Mental Health Weekly Digest*, p. 31.

Navarrete, C. D., & Fessler, D. M. T. (2006). Disease avoidance and ethnocentrism: The effects of disease vulnerability and disgust sensitivity on intergroup attitudes. *Evolution and Human Behavior*, *27*, 270–282.

Neblo, M. A. (2007). Philosophical psychology with political intent. In R. Neuman, M. McKuen, A. Crigler, & G. Marcus (Eds.), *The affect effect: Dynamics of emotion in political thinking and behavior* (pp. 25–47). Chicago, IL: University of Chicago Press.

New, A. S., Hazlett, E. A., Buchsbaum, M. S., Goodman, M., Mitelman, S. A., Newmark, R., & Siever, L. J. (2007). Amygdala_prefrontal disconnection in borderline personality disorder. *Neuropsychopharmecology*, *32*, 1629–1640.

Niedenzu, H.-J., Meleghy, T., & Meyer, P. (Eds.). (2008). *The new evolutionary social science*. Boulder, CO: Paradigm Publishers.

Oakeshott, M. (1962). *Rationalism in politics*. New York, NY: Basic Books.

O'Donovan, M. C., & Owen, M. J. (2009). Genetics and the brain: Many pathways to enlightenment. *Human Genetics*, *126*(1), 91–99.

Odum, E. (1971). *Fundamentals of ecology* (3rd ed.). Philadelphia, PA: Saunders.

Olatunji, B. O. (2008). Disgust, scrupulosity and conservative attitudes about sex: Evidence for a mediational model of homophobia. *Journal of Research in Personality*, *42*(5), 1364–1369.

Olatunji, B. O., Adams, T., Ciesielski, B., David, B., Sarawgi, S., & Broman-Fulks, J. (2012). The three domains of disgust scale factor structure, psychometric properties, and conceptual limitations. *Assessment*, *19*(2), 205–225.

Olsson, A., Nearing, K. I., & Phelps, E. A. (2007). Learning fears by observing others: The neural systems of social fear transmission. *Social Cognitive and Affective Neuroscience*, *2*(1), 3–11.

Ophuls, W. (1977). *Ecology and the politics of scarcity*. San Francisco, CA: W. H. Freeman and Company.

Ophuls, W. (1997). *Requiem for modern politics*. Boulder, CO: Westview Press.

Oskarsson, S., Dawes, C. T., Johannesson, M., & Magnusson, P. K. E. (2012). The genetic origins of the relationship between psychological traits and social trust. *Twin Research and Human Genetics, 15, 92*(1), 21–33.

Ostrom, E. (1998). A behavioral approach to the rational choice theory of collective action: Presidential address, American political science association, 1997. *American Political Science Review*, *92*(1), 1–22.

Ostrom, E. (2011). Background on the institutional analysis and development framework. *Policy Studies Journal*, *39*(1), 7–27.

O'Sullivan, F., & Paltsev, S. (2012). *Environmental research letters* 7 044030. Retrieved from http://iopscience.iop.org/1748-9326/7/4/044030/pdf/1748-9326_7_4_044030.pdf. Accessed on November 26.

Oxley, D., Smith, K. B., Alford, J. R., Hibbing, M. V., Miller, J. L., Scalora, M., … Hibbing, J. R. (2008). Political attitudes vary with physiological traits. *Science*, *321*(5896), 1667–1670.

Page, B. I., & Shapiro, R. Y. (2010). *The rational public: Fifty years of trends in Americans' policy preferences*. Chicago, IL: University of Chicago Press.

Panksepp, J. (1998). *Affective neuroscience: The foundations of human and animal emotions*. New York, NY: Oxford University Press.

Pantic, M., Cowie, R., D'Errico, F., Heylen, D., Mehu, M., Pelachaud, C., & Vinciarelli, A. (2011). Social signal processing: The research agenda. In T. B. Moeslund, A. Hilton, V. Krüger, & L. Sigal (Eds.), *Visual analysis of humans* (pp. 511–538). New York, NY: Springer.

Park, B. E. (1986). *The impact of illness on world leaders*. Philadelphia, PA: University of Pennsylvania Press.

Park, B. E. (1988). Presidential disability: Past experiences and future implications. *Politics and the Life Sciences*, *7*(1), 50–66.

Penney, D., & McGee, G. (2005). Chemical trust: Oxytocin oxymoron? *The American Journal of Bioethics*, *5*(3), 1–2.

Perry, B. D. (2002). Childhood experience and the expression of genetic potential: What childhood neglect tells us about nature and nurture. *Brain and Mind*, *3*, 79–100.

Pessoa, L., & Adolphs, R. (2010). Emotion processing and the amygdala: From a "low road" to "many roads" of evaluating biological significance. *Nature Reviews Neuroscience*, *11*, 773–783.

Pessoa, L., & Engelmann, J. B. (2010). Embedding reward signals into perception and cognition. *Frontiers in Neuroscience*, *4*, 1–8.

Pessoa, L. (2009). How do emotion and motivation direct executive control? *Trends in Cognitive Science*, *13*(4), 160–166.

Pessoa, L. (2010a). Emergent processes in cognitive-emotional interactions. *Dialogues in Clinical Neuroscience*, *12*(4), 433–448.

Pessoa, L. (2010b). Emotion and cognition and the amygdala: From "what is it?" to "what's to be done?" *Neuropsychologia*, *48*(12), 3416–3429.

Petersen, M. B., & Aarøe, L. (2013). Politics in the mind's eye: Imagination as a link between social and political cognition. *American Political Science Review*, *107*(2), 275–293.

Petersen, M. B., & Kennair, L. E. O. (2009). Evolutionary political science – Integrating evolutionary psychology and the social sciences. In H. Høgh-Olesen, P. Bertelsen, & J. Tønnesvang (Eds.), *Human characteristics* (pp. 313–350). Cambridge: Cambridge Scholars Publishing.

Petersen, M. B. (2010). Distinct emotions, distinct domains: Anger, anxiety and perceptions of intentionality. *The Journal of Politics*, *72*(2), 357–365.

Petersen, M. B., Sznycer, D., Cosmides, L., & Tooby, J. (2012). Who deserves help? Evolutionary psychology, social emotions, and public opinion about welfare. *Political Psychology*, *33*(3), 395–418.

Petersen, M. B., Sznycer, D., Sell, A., Cosmides, L., & Tooby, J. (2013). The ancestral logic of politics: Upper-body strength regulates men's assertion of self-interest over economic redistribution. *Psychological Science*, *24*(7), 1098–1103.

Peterson, S. A., & Lawson, R. (1989). Risky business: Prospect theory and politics. *Political Psychology*, *10*(2), 325–339.

Peterson, S. A., & Somit, A. (1982). Methodological problems associated with a biologically oriented social science. In T. C. Wiegele (Ed.), *Biology and the social sciences: An emerging revolution* (pp. 349–366). Boulder, CO: Westview Press.

Peterson, S. A., & Somit, A. (1989). Primates, coalitions, and small group politics. *Commonwealth*, *3*, 18–33.

Peterson, S. A., & Somit, A. (Eds.). (2013). *The world of biology and politics: Organization and research areas* (Vol. 11). Research in Biopolitics. Bingley, UK: Emerald Group Publishing Limited.

Peterson, S. A. (1987). Biosocial predictors of older Americans' political participation. *Politics and the Life Sciences*, *3*, 246–251.

Peterson, S. A. (1990). *Political behavior: Patterns in everyday life*. Newbury Park, CA: Sage Publications.

Peterson, S. A. (1991). The health of the polity: Results from a national survey. *Politics and the Life Sciences*, *10*(1), 45–56.

Peterson, S. A. (2013). Toward an evolutionary synthesis of individualist and communist anarchism: The evolutionary bases of cooperation. In S. A. Peterson & A. Somit (Eds.), *The world of biology and politics: Organization and research areas* (Vol. 11, pp. 185–198). Research in Biopolitics. Bingley, UK: Emerald Group Publishing Limited.

Petron, G., Frost, G. J., Trainer, M. K., Miller, B. R., Dlugokencky, E. J., & Tans, P. (2011). Hydrocarbon emissions characterization in the Colorado front range – A pilot study. *Journal of Geophysical Research*, doi:10.1029/2011JD01636.

Pettman, R. (1975). *Human behavior and world politics*. New York, NY: St. Martin's Press.

Pettman, R. (1981). *Biopolitics and international values: Investigating liberal norms*. New York, NY: Pergamon Press.

Pfaff, W. (1993). *The wrath of nations: Civilization and the furies of nationalism*. New York, NY: Simon and Schuster.

Philips, C. (1999). Political scientists need to pay attention to developments in biology. *Political Science & Politics*, *32*(3), 535–536.

Piketty, T. (2014). *Capital in the twenty first century*. Cambridge, MA: The Belknap Press of Harvard University.

Pinker, S. (2002). *The blank slate: The modern denial of human nature*. New York, NY: Viking.

Pinker, S. (2011). *The better angels of our nature: Why violence has declined*. New York, NY: Viking.

Pirages, D. C., & DeGeest, T. M. (2004). *Ecological security: An evolutionary perspective on globalization*. Lanham, MD: Rowman and Littlefield.

Pirages, D. (1978). *Global ecopolitics: A new context for international relations*. North Scituate, MA: Duxbury Press.

Poteete, A. R., Janssen, M. A., & Ostrom, E. (2010). *Working together: Collective action, the commons, and multiple methods in practice*. Princeton, NJ: Princeton University Press.

Poutvaara, P., Jordahl, H., & Berggren, N. (2009). Faces of politicians: Babyfacedness predicts inferred competence but not electoral success. *Journal of Experimental Social Psychology*, *45*(5), 1132–1135.

Preston, S. (2011). The empathy gap. *Nature*, *472*, 416.

Price, J. L., & Drevets, W. C. (2010). Neurocircuitry of mood disorders. *Neuropsychopharmacology*, *35*(1), 192–216.

Prindle, D. F. (2012). Importing concepts from biology into political science: The case of punctuated equilibrium. *Policy Studies Journal*, *40*(1), 21–43.

Quattrone, G. A., & Tversky, A. (1988). Contrasting rational and psychological analyses of political choice. *American Political Science Review*, *82*(3), 719–736.

Raafat, R. M., Chater, N., & Firth, C. (2009). Herding in humans. *Trends in Cognitive Science*, *13*(10), 420–428.

Raine, A., & Yang, Y. (2006). Neural foundations to moral reasoning and antisocial behavior. *Social Cognitive and Affective Neuroscience*, *1*(3), 203–213.

Raine, A. (2008). From genes to brain to antisocial behavior. *Current Directions in Psychological Science*, *17*, 323–328.

Reeves, R. R., & Panguluri, R. L. (2011). Neuropsychiatric complications of traumatic brain injury. *Journal of Psychological Nursing and Mental Health Services*, *49*(3), 42–51.

Rich, A. L., & Crosby, E. C. (2013). Analysis of reserve pit sludge from unconventional natural gas hydraulic fracturing and drilling operations for the presence of technologically enhanced naturally occurring radioactive material (TENORM). *New Solutions*, *23*(1), 117–135. Retrieved from http://baywood.metapress.com/media/lhe1qpmuml7qhfpuupvn/contributions/k/6/2/1/k621376330557386_html/fulltext.html

Richardson, R. C. (2007). *Evolutionary psychology as maladapted psychology*. Cambridge, MA: MIT Press.

Ridley, M. (2010). *The rational optimist: How prosperity evolves*. New York, NY: Harper.

Riemann, R., Angleitner, A., & Strelau, J. (1997). Genetic and environmental influences on personality: A study of twins reared together using the self- and peer report NEO-FFI scales. *Journal of Personality*, *65*(3), 449–475.

Roberts, M. (1938). *Biopolitics: An essay in physiology, pathology and the politics of social and somatic organisms*. London: J. M. Dent.

Robins, R. S., & Rothschild, H. (1981). Hidden health disabilities and the presidency: Medical management and political considerations. *Perspectives in Biology and Medicine*, *24*, 241–253.

Robins, R. S., & Rothschild, H. (1988). Ethical dilemmas of the president's physician. *Politics and the Life Sciences*, *7*(1), 3–11.

Rodrigues, S. M., LeDoux, J. E., & Sapolsky, R. M. (2009). The influence of stress hormones on fear circuitry. *Annual Review of Neuroscience*, *32*, 289–313.

Rule, N. O., Freeman, J. B., Moran, J. M., Gabrieli, J. D. E., Adams, R. B., & Ambady, N. (2010). Voting behavior is reflected in amygdala response across cultures. *Social Cognitive and Affective Neuroscience*, *5*(2–3), 349–355.

Russell, P. S., & Giner-Sorolla, R. (2013). Bodily moral disgust: What it is, how it is different from anger, and why it is an unreasoned emotion. *Psychological Bulletin*, *139*(2), 328–351.

Rustow, D., & Erickson, K. (Eds.). (1991). *Comparative political dynamics: Global research perspectives*. New York, NY: Harper Collins.

Saad, G. (2007). *The evolutionary bases of consumption*. Mahwah, NJ: Lawrence Erlbaum.

Saad, G. (2011a). *The consuming instinct: What juicy burgers, ferraris, pornography, and gift giving reveal about human nature*. New York, NY: Prometheus Books.

Saad, G. (2011b). *Evolutionary psychology in the business sciences*. New York, NY: Springer.

Sachs, J. D. (2008). *Common wealth: Economics for a crowded planet*. New York, NY: Penguin Press.

Salter, F. K. (1996). "Drawn by the light": Visual recording methods in biopolitics. In A. Somit & S. A. Peterson (Eds.), *Research in biopolitics* (Vol. 4, pp. 23–60). Greenwich, CT: JAI Press.

Salter, F. K. (2000). The Universals of Command: Biological Constraints on Institutional Structures. Paper presented at annual meeting of the Association for Politics and the Life Sciences, Washington, DC, September.

Salter, F. K. (2002). *Risky transactions: Trust, kinship, and ethnicity*. Oxford, NY: Berghahn Books.

Salter, F. K. (2007). *Emotions in command: Biology, bureaucracy, and cultural evolution*. New Brunswick, NJ: Transaction Publications.

Salzman, C. D., & Fusi, S. (2010). Emotion, cognition, and mental state representation in amygdala and prefrontal cortex. *Annual Review of Neuroscience*, *33*, 173–202.

Sanders, G. S. (1982). Social comparison and perceptions of health and illness. In G. S. Sanders (Ed.), *Social psychology of health and illness*. Hillsdale, NJ: Lawrence Erlbaum.

Sanderson, S. K. (2001). *The evolution of human sociality*. Lanham, MD: Rowman & Littlefield Publishers.

Sanderson, S. K. (2008). The impact of Darwinism on sociology: An historical and critical view. In H.-J. Niedenzu, T. Meleghy, & P. Meyer (Eds.), *The new evolutionary social science* (pp. 29–41). Boulder, CO: Paradigm Publishers.

Sanderson, S. K. (2014). *Human nature and the evolution of society*. Boulder, CO: Westview Press.

Sanfey, A. G. (2009). Expectations and social decision-making: Biasing effects of prior knowledge on ultimatum responses. *Mind and Society*, *8*, 93–107.

Sato, W., Kochiyama, T., Matsuda, K., Usui, K., Inoue, Y., & Toichi, M. (2011). Rapid amygdala gamma oscillations in response to fearful facial expressions. *Neuropsychologia*, *49*, 612–617.

Schaich Borg, J., Lieberman, D., & Kiehl, K. A. (2008). Infection, incest, and iniquity: Investigating the neural correlates of disgust and morality. *Journal of Cognitive Neuroscience*, *20*(9), 1529–1546.

Schaller, M., & Murray, D. R. (2011). Infectious disease and the creation of culture. *Advances in Culture and Psychology*, *1*, 99–151.

Schaller, M., & Park, J. H. (2011). The behavioral immune system (and why it matters). *Current Directions in Psychological Science*, *20*(2), 99–103.

Schlozman, K. L. (2013). Two concerns about ten misconceptions. *Perspectives on Politics*, *11*(2), 490–491.

Schmook, M. T., Brugger, P. C., Weber, M., Kasprian, G., Nemec, S., Krampl-Bettelheim, E., & Prayer, D. (2010). Forebrain development in fetal MRI: Evaluation of anatomical landmarks before gestational week 27. *Neuroradiology Heidelberg*, *52*(6), 495–499.

Schoen, H., & Steinbrecher, M. (2013). Beyond total effects: Exploring the interplay of personality and attitudes in affecting turnout in the 2009 German federal election. *Political Psychology*, *34*(4), 533–552. doi:10.1111/pops.12031

Schoenbaum, G., Saddoris, M. P., & Stalnaker, T. A. (2007). Reconciling the roles of orbitofrontal cortex in reversal learning and the encoding of outcome expectancies. *Annals of the New York Academy of Sciences*, *1121*, 320–335.

Schreiber, D. (2005). *Evaluating politics: A search for the neural substrates of political thought*. PhD Dissertation, University of California, Los Angeles.

Schreiber, D. (2011). From SCAN to neuropolitics. In P. K. Hatemi & R. McDermott (Eds.), *Man is by nature a political animal*. Chicago, IL: University of Chicago Press.

Schreiber, D., Fonzo, G., Simmons, A. N., Dawes, C. T., Flagan, T., Fowler, J. H., & Paulus, M. P. (2013). Red brain, blue brain: Evaluative processes differ in democrats and republicans. PloS One, *8*(2), e52970. doi:10.1371/journal.pone.0052970

Schubert, G. (1976). Politics as a life science: How and why the impact of modern biology will revolutionize the study of political behavior. In A. Somit (Ed.), *Biology and politics* (pp. 155–198). The Hague: Mouton.

Schubert, J. (1981). The impact of food aid on world malnutrition. *International Organization*, *35*, 329–354.

Schubert, G. (1982). Politics as a life science. In T. C. Wiegele (Ed.), Biology and the *s*ocial sciences: An emerging revolution. Boulder, CO: Westview Press.

Schubert, G. (1985). *Political culture and judicial behavior*. Lanham, MD: University Press of America.

Schubert, J. N. (1988). Age and active-passive leadership style. *American Political Science Review*, *82*(3), 763–772.

Schubert, G. (1989). *Evolutionary politics*. Carbondale, IL: Southern Illinois University Press.

Schubert, J. (1996). Verbal, vocal, and visual aspects of political speech: A case study of Clinton's inaugural speech. In A. Somit & S. A. Peterson (Eds.), *Research in biopolitics* (Vol. 4, pp. 113–138). London: JAI Press.

Schubert, G. (1997). Oral argument in Supreme Court decision making: Ethological and paralinguistic analysis of psychophysiological biosocial political behavior. In A. Somit & S. Peterson (Eds.), *Research in biopolitics* (Vol. 5, pp. 3–56). London: JAI Press.

Schubert, J. N., Curran, M. A., & Strungaru, C. (2011). Physical attractiveness, issue agreement, and assimilation effects in candidate appraisal. *Politics and the Life Sciences*, *30*(1), 33–49.

Schubert, J. N., Wiegele, T. C., & Hines, S. M. (1987). Age and political behavior in collective decision-making. *International Political Science Review*, *8*(2), 131–146.

Schubert, J., Curran, M., & Strungaru, C. X. (2000). Appearance effects in candidate appraisal: Facial averageness, symmetry, dominance, and electability for male and female candidates. Paper presented at annual meeting of the Association for Politics and the Life Sciences, Washington, DC, September.

Schubert, J. N., Peterson, S. A., Schubert, G., & Wasby, S. (1992). Observing supreme court oral argument: A biosocial approach. *Politics and the Life Sciences*, *11*(1), 35–51.

Schubert, J. N., Stewart, P. A., & Curran, M. A. (2002). A defining presidential moment: 9/11 and the rally effect. *Political Psychology*, *23*(3), 559–583.

Schubert, J., Wiegele, T., & Hines, S. M. (1986). Report to national institute of aging. In *Age, age structure, and political decision making*. DeKalb, IL: Program for Biosocial Research.

Schubert, J., Wiegele, T., & Hines, S. M. (1987). Age and political behavior in collective decision-making. *International Political Science Review*, *8*(2), 131–146.

Schumann, C. M., Bauman, M. D., & Amaral, D. G. (2011). Abnormal structure or function of the amygdale is a common component of neurodevelopmental disorders. *Neuropsychologia*, *49*, 745–759.

Schwartz, D. C. (1976). Somatic states and political behavior. In A. Somit (Ed.), *Biology and politics*. Paris: Mouton.

Sears, D. O. (1993). Symbolic politics: A socio-psychological theory. In S. Iyengar & W. J. McGuire (Eds.), *Explorations in political psychology* (pp. 113–149). Durham, NC: Duke University Press.

Segerstrale, U. (1998). Nonverbal communication: Persuading the social sciences that biology doesn't bite. Paper presented at annual meeting of the Association for Politics and the Life Sciences, Boston, August 31.

Seligman, R., & Brown, R. A. (2010). Theory and method at the intersection of anthropology and cultural neuroscience. *Social Cognitive and Affective Neuroscience*, *5*(2–3), 130–137.

Sell, A., Cosmides, L., Tooby, J., Sznycer, D., Von Rueden, C., & Gurven, M. (2009). Human adaptations for the visual assessment of strength and fighting ability from the body and face. *Proceedings of the Royal Society B: Biological Sciences*, *276*(1656), 575–584.

Seo, D., Patrick, C. J., & Kennealy, P. J. (2008). Role of serotonin and dopamine system interactions in the neurobiology of impulsive aggression and its comorbidity with other clinical disorders. *Aggression and Violent Behavior*, *13*, 383–395.

Settle, J. E., Dawes, C. T., & Fowler, J. H. (2009). The heritability of partisan attachment. *Political Research Quarterly*, *62*(3), 601–613.

Shadish, W. R., Cook, T. D., & Campbell, D. T. (2002). *Experimental and quasi-experimental designs for generalized causal inference*. Boston, MA: Houghton Mifflin Company.

Shaw, R., & Wong, Y. (1989). *Genetic seeds of warfare*. Boston, MA: Unwin Hyman.

Shcherbak, I., Millar, N., & Robertson, G. P. (2014). *Global metaanalysis of the nonlinear response of soil nitrous oxide (N2O) emissions to fertilizer nitrogen*. PNAS Early Edition. Retrieved from www.pnas.org/content/early/2014/06/04/1322434111.full.pdf +html?sid=6535a3dc-edbb-44e6-af5a-f43a73a5c64a. Accessed on April 6.

Sherman, G. D., Lee, J. J., Cuddy, A. J., Renshon, J., Oveis, C., Gross, J. J., & Lerner, J. S. (2012). Leadership is associated with lower levels of stress. *Proceedings of the National Academy of Sciences of the United States of America*, *109*(44), 17903–17907.

Shirtcliff, E. A., Vitacco, M. J., Graf, A. R., Gostisha, A. J., Merz, J. L., & Zahn-Waxler, C. (2009). Neurobiology of empathy and callousness: Implications for the development of antisocial behavior. *Behavioral Sciences and the Law*, *27*, 137–171.

Shultziner, D. (2013). Genes and politics: A new explanation and evaluation of twin study results and association studies in political science. *Political Analysis*, *21*, 350–367.

Sibley, M. Q. (1972). The relevance of classical political theory for economy, technology. Paper at the 1972 Annual Meeting of the American Political Science Association, Washington, DC, September 5–9.

Sibley, M. Q. (1977). *Nature and civilization: Some implications for politics*. Itasca, IL: F.E. Peacock Publishers, Inc.

Siever, L. J. (2008). Neurobiology of aggression and violence. *American Journal of Psychiatry*, *165*, 429–442.

Sigelman, L. (2006). Report of the editor of the American political science review, 2004–2005. *PS: Political Science and Politics*, *39*, 171–173.

Simon, H. (1983). *Models of bounded rationality*. Cambridge, MA: MIT Press.

Singer, T., & Lamm, C. (2009). The social neuroscience of empathy. *Annals of the New York Academy of Sciences*, *1156*, 81–96.

Singer, T., Seymour, B., O'Doherty, J. P., Kaube, H., Dolan, R. J., & Frith, C. D. (2004). Empathy for pain involves the affective but not sensory components of pain. *Science*, *303*(5661), 1157–1162.

Singer, T., Seymour, B., O'Doherty, J. P., Stephan, K. E., Dolan, R. J., & Frith, C. D. (2006). Empathic neural responses are modulated by the perceived fairness of others. *Nature, 439*, 466–469.

Siverman, I., & Fisher, M. L. (2001). Is psychology undergoing a paradigm shift? Past, present, and future roles of evolutionary psychology. In A. Somit (Ed.), *Evolutionary approaches in the behavioral sciences: Toward a better understanding of human nature* (pp. 203–216). Greenwich, CT: JAI Press.

Skitka, L. J., Bauman, C. W., & Mullen, E. (2004). Political tolerance and coming to psychological closure following the September 11, 2001, terrorist attacks: An integrative approach. *Personality & Social Psychology Bulletin, 30*(6), 743–756.

Smith, K. B., & Hibbing, J. R. (2011). The mind body connection: Psychophysiology as an approach to studying political attitudes and behaviors. In P. K. Hatemi & R. McDermott (Eds.), *Man is by nature a political animal: Evolution, biology, and politics*. Chicago, IL: University of Chicago Press.

Smith, K., Alford, J. R., Hatemi, P. K., Eaves, L. J., Funk, C., & Hibbing, J. R. (2012). Biology, ideology, and epistemology: How do we know political attitudes are inherited and why should we care? *American Journal of Political Science, 56*, 17–33.

Smith, K. B., Larimer, C. W., Littvay, L., & Hibbing, J. R. (2007). Evolutionary theory and political leadership: Why certain people do not trust decision makers. *Journal of Politics, 69*(2), 286–299.

Smith, K. B., Oxley, D., Hibbing, M. V., Alford, J. R., & Hibbing, J. R. (2011a). Disgust sensitivity and the neurophysiology of left-right political orientations. PloS One, *6*(10), e25552.

Smith, K. B., Oxley, D. R., Hibbing, M. V., Alford, J. R., & Hibbing, J. R. (2011b). Linking genetics and political attitudes: Reconceptualizing political ideology. *Political Psychology, 32*(3), 369–397.

Snow, C. P. (1990). The two cultures. *Leonardo, 23*(2/3), 169–173.

Sokolon, M. (2006). *Political emotions: Aristotle and the symphony of reason and emotion*. DeKalb, IL: Northern Illinois University Press.

Soliman, A., Bagby, R. M., Wilson, A. A., Miler, L., Clark, M., Rusjan, P., … Meyer, J. H. (2010). Relationship of monoamine oxidase A binding to adaptive and maladaptive personality traits. *Psychological Medicine, 1*, 1–10.

Solomon, S. (2010). *Water: The epic struggle for wealth, power, and civilization*. New York, NY: Harper Collins.

Somit, A., & Peterson, S. A. (Eds.). (1989). *The dynamics of evolution: The punctuated equilibrium debate in the natural and social sciences*. Ithaca, NY: Cornell University Press.

Somit, A., & Peterson, S. A. (1995). Darwinism, dominance, and democracy. In J. S. Losco (Ed.), *Human nature in politics* (pp. 19–34). Greenwich, CT: JAI Press.

Somit, A., & Peterson, S. A. (1997). *Darwinism, dominance, and democracy*. Westport, CT: Praeger.

Somit, A., & Peterson, S. A. (1998). Review article: Biopolitics after three decades – A balance sheet. *British Journal of Politics, 28*, 559–571.

Somit, A., & Peterson, S. A. (1999). Rational choice and biopolitics: A (Darwinian) tale of two theories. *PS: Political Science and Politics, 32*(1), 39–44.

Somit, A., & Peterson, S. A. (2000). Review article: Biopolitics after three decades – A balance sheet. *British Journal of Political Science, 28*, 559–571.

Somit, A., & Peterson, S. A. (2003b). From human nature to public policy: Evolutionary theory challenges the "standard model". In A. Somit & S. A. Peterson (Eds.), *Human nature and public policy: An evolutionary approach* (pp. 3–18). New York, NY: Palgrave Macmillan.

Somit, A., & Peterson, S. A. (2005). *The failure of democratic nation building: Ideology meets evolution*. New York, NY: Palgrave Macmillan.

Somit, A., & Peterson, S. A. (2013b). Biology and politics: An introduction. In S. A. Peterson, & A. Somit (Eds.), *The world of biology and politics: Organization and research areas* (Vol. 11, pp. 1–11). Research in Biopolitics. Bingley, UK: Emerald Group Publishing Limited.

Somit, A., & Watts, M. (1994). Biology and politics: Political, practical, and philosophical issues. In A. Somit & S. A. Peterson (Eds.), *Research in biopolitics* (Vol. 2). London: JAI Press.

Somit, A. (1968). Toward a more biologically oriented political science: Ethology and psychopharmacology. *Midwest Journal of Political Science, 12*, 550–567.

Somit, A. (1972). Biopolitics. *British Journal of Political Science, 2*, 209–238.

Somit, A. (1976). *Biology and politics: Recent explorations*. Paris: Mouton.

Somit, A. (1981). Human nature as the central issue in political philosophy. In E. White (Ed.), *Sociobiology and human nature* (pp. 167–180). Lexington, MA: D. C. Heath.

Somit, A. (2011). What hath biopolitics wrought? *Politics and the Life Sciences, 30*(1), 95–97.

Somit, A., & Peterson, S. A. (2013a). International political science association's research Committee #. In S. A. Peterson & A. Somit (Eds.), *The world of biology and politics: Organization and research areas*. (Vol. 11, pp. 15–19). Research in Biopolitics. Bingley, UK: Emerald Group Publishing Limited.

Somit, A., Arwine, A., & Peterson, S. A. (1996). *Birth order and political behavior*. Lanham, MD: University Press of America.

Somit, A., Arwine, A., & Peterson, S. A. (1997). Birth order and female federal judges. In A. Somit & S. A. Peterson (Eds.), *Research in biopolitics* (Vol. 5, pp. 103–108). London: JAI Press.

Somit, A., Seo, K.-K., & Peterson, S. A. (1994). The impact of biology on the study of international politics. In A. Somit & S. A. Peterson (Eds.), *Biopolitics in the mainstream*. (Vol. 2, pp. 135–140). Research in Biopolitics.London: JAI Press.

Spence, M. (2002). Signaling in retrospect and the informational structure of markets. *The American Economic Review, 92*(3), 434–459.

Spezio, M. L., Rangel, A., Alvarez, R. M., O'Doherty, J. P., Mattes, K., Todorov, A., … Adolphs, R. (2008). A neural basis for the effect of candidate appearance on election outcomes. *Social, Cognitive and Affective Neuroscience, 3*, 344–352.

Spisak, B. R., Dekker, P. H., Krüger, M., & van Vugt, M. (2012). Warriors and peacekeepers: Testing a biosocial implicit leadership hypothesis of intergroup relations using masculine and feminine faces. PloS One, *7*(1), e30399.

Spisak, B. R., Nicholson, N., & van Vugt, M. (2011). Leadership in organizations: An evolutionary perspective. In G. Saad (Ed.), *Evolutionary psychology in the business sciences* (pp. 165–190). New York, NY: Springer.

Spisak, B. R., Homan, A. C., Grabo, A., & Van Vugt, M. (2012). Facing the situation: Testing a biosocial contingency model of leadership in intergroup relations using masculine and feminine faces. *The Leadership Quarterly, 23*(2), 273–280.

SPM. (2014). Summary for Policymakers (SPM) (2014). Retrieved from http://www.climatechange2013.org/images/report/WG1AR5_SPM_FINAL.pdf

Spretnak, C., & Capra, F. (1985). *Green politics*. London: Paladin.

Sprout, H., & Sprout, M. (1965). *The ecological perspective on human affairs with special reference to international politics*. Princeton, NJ: Princeton University Press.

Sprout, H., & Sprout, M. (1968). *An ecological paradigm for the study of international politics*. Princeton, NJ: Princeton Center for International Studies.

Stam, A. C., Von Hagen-Jamar, A., & Worthington, A. B. H. (2012). Fear and attitudes towards torture and preventive war. *Twin Research and Human Genetics*, *15*, 60–70.

Stanton, S. J., Beehner, J. C., Saini, E. K., Kuhn, C. M., & LaBar, K. S. (2009). Dominance, politics, and physiology: Voters' testosterone changes on the night of the 2008 United States presidential election. PloS One, *4*(10), e7543. doi:10.1371/journal.pone.0007543

Stavrakakis, Y. (1997). Green ideology: A discursive reading. *Journal of Political Ideologies*, *2*(3), 259–279.

Stedman, R. C., Jacquet, J. B., Filteau, M. R., Willits, F. K., Brasier, K. J., & McLaughlin, D. K. (2012). Marcellus shale gas development and new boomtown research: Views of New York and Pennsylvania residents. *Journal of Environmental Practice*, *14*, 287–298.

Stevens, D., Sullivan, J., Allen, B., & Alger, D. (2008). What's good for the goose is bad for the gander: Negative political advertising, partisanship, and turnout. *Journal of Politics*, *70*, 527–541.

Stewart, P. A. (2008). Subliminals in the 2000 presidential election: Policy implications of applied neuroscience. *Public Integrity*, *10*(3), 215–232.

Stewart, P. A. (2012). *Debatable humor: Laughing matters on the 2008 presidential primary campaign*. Lanham, MD: Lexington Books.

Stewart, P. A. (2013). The two cultures of "biopolitics": Comment in response to Liesen and Walsh. *Politics and the Life Sciences*, *32*(1), 94–98.

Stewart, P. A., Salter, F. K., & Mehu, M. (2009). Taking leaders at face value: Ethology and the analysis of televised leader displays. *Politics and the Life Sciences*, *28*(1), 48–74.

Stone, D. A. (1997). *Policy paradox: The art of political decision making*. New York, NY: W. W. Norton.

Strate, J. M. (1991). Health and political participation: A neglected topic. *Politics and the Life Sciences*, *10*(1), 63–64.

Sturgis, P., Read, S., Hatemi, P., Zhu, G., Trull, T., Wright, M., & Martin, N. (2010). A genetic basis for social trust? *Political Behavior*, *32*(2), 205–230.

Sueur, C., Jacobs, A., Amblard, F., Petit, O., & King, A. J. (2011). How can social network analysis improve the study of primate behavior? *American Journal of Primatology*, *73*(8), 703–719.

Sullivan, D. G., & Masters, R. D. (1994). Biopolitics, the media, and leadership: Nonverbal cues, emotions, and trait attributions in the evaluation of leaders. In A. Somit & S. A. Peterson (Eds.), *Research in biopolitics* (Vol. 2, pp. 237–273). London: JAI Press.

Sutherland, J. (1973). *A general systems philosophy for the social and the behavioral sciences*. New York, NY: George Braziller.

Svolik, M. (2008). Authoritarian reversals and democratic consolidation. *American Political Science Review*, *102*(2), 153–168.

Tancredi, L. (2005). *Hardwired behavior: What neuroscience reveals about morality*. Cambridge, UK: Cambridge University Press.

Terrizzi, J. A., Jr., Shook, N. J., & McDaniel, M. A. (2012). The behavioral immune system and social conservatism: A meta-analysis. *Evolution and Human Behavior*, *34*, 99–108.

Terrizzi, J. A., Shook, N. J., & Ventis, W. L. (2010). Disgust: A predictor of social conservatism and prejudicial attitudes toward homosexuals. *Personality and Individual Differences*, *49*(6), 587–592.

Thayer, B. A. (2004). *Darwin and international relations: On the evolutionary origins of war and conflict*. Lexington, KY: University Press of Kentucky.

The White House. (2013). *Brain initiative*. Retrieved from http://www.whitehouse.gov/share/brain-initiative; http://www.whitehouse.gov/blog/2013/04/02/brain-initiative-challenges-researchers-unlock-mysteries-human-mind

Theodoridis, A. G., & Nelson, A. J. (2012). Of BOLD claims and excessive fears: A call for caution and patience regarding political neuroscience. *Political Psychology*, *33*(1). doi:10.1111/j.1467-9221.2011.00860.x

Thompson, W. R. (Ed.). (2001). *Evolutionary interpretations of world politics*. New York, NY: Routledge.

Thorson, T. L. (1970). *Biopolitics*. New York, NY: Holt, Rinehart and Winston.

Thorson, T. L. (1982). Review of biology and politics. *Politics and the Life Sciences*, *1*(1), 71–73.

Tiger, L., & Fox, R. (1971). *The imperial animal*. Brunswick, NJ: Transaction Publishers.

Tinbergen, N. (1963). On aims and methods of ethology. *Zeitschrift für Tierpsychologie*, *20*(4), 410–433.

Tingley, D. (2006). Neurological imaging as evidence in political science: A review, critique, and guiding assessment. *Social Sciences Information. Information Sur les Sciences Sociales*, *45*(1), 5–33.

Tingley, D. (2007). Evolving political science: Biological adaptation, rational action, and symbolism. *Politics and the Life Sciences*, *25*(1/2), 23–41.

Todorov, A., Mandisodza, A. N., Goren, A., & Hall, C. C. (2005). Inferences of competence from faces predict election outcomes. *Science*, *308*, 1623–1626.

Todorov, A., Pakrashi, M., & Oosterhof, N. N. (2009). Evaluating faces on trustworthiness after minimal time exposure. *Social Cognition*, *27*(6), 813–833.

Tollefson, J. (2013). Methane leaks erode green credentials of natural gas. *Nature*, *493*, 12.

Tooby, J., & Cosmides, L. (1988). *The evolution of war and its cognitive foundation*. Technical Report 88-1. Institute for Evolutionary Studies.

Tooby, J., & Cosmides, L. (1992). The psychological foundations of culture. In J. Barkow, L. Cosmides, & J. Tooby (Eds.), *The adapted mind: Evolutionary psychology and the generation of culture* (pp. 19–136). New York, NY: Oxford University Press.

Toulmin, S. (1991). *Cosmopolis: The hidden agenda of modernity*. New York, NY: Free Press.

Trivers, R. L., & Willard, D. E. (1973). Natural selection of parental ability to vary the sex ratio of offspring. *Science*, *179*, 90–92.

Trivers, R. L. (1971). The evolution of reciprocal altruism. *Quarterly Review of Biology*, *46*, 35–57.

Turner, J. H. (1998). *The structure of sociological theory*. Belmont, CA: Wadsworth.

Tuschman, A. (2013). *Our political nature: The evolutionary origins of what divides us*. New York, NY: Prometheus.

Tybur, J. M., Lieberman, D., & Griskevicius, V. (2009). Microbes, mating, and morality: Individual differences in three functional domains of disgust. *Journal of Personality and Social Psychology*, *97*(1), 103–122.

Tybur, J. M., Lieberman, D., Kurzban, R., & DeScioli, P. (2012). Disgust: Evolved function and structure. *Psychological Review*, *120*(1), 65–84.

Tybur, J., Merriman, L., Caldwell Hooper, A., McDonald, M., & Navarrete, C. (2010). Extending the behavioral immune system to political psychology: Are political conservatism and disgust sensitivity really related. *Evolutionary Psychology*, *8*, 599–616.

US Geographical Survey (USGS). (2013). *FAQs*. Retrieved from http://www.usgs.gov/faq/index.php?sid=54684&lang=en&action=artikel&cat=229&id=2292&artlang=en

Van Bavel, J. J., Packer, D. J., & Cunningham, W. A. (2008). The neural substrates of in-group bias: A functional magnetic resonance imaging investigation. *Psychological Science*, *19*(11), 1131–1139.

Van den Berghe, P. (1981). *The ethnic phenomenon*. New York, NY: Elsevier.

Van der Dennen, J. M. G. (1995). *The origin of war: The evolution of a male-coalitional male reproductive strategy*. Groningen: Origin Press.

Van der Dennen, J. M. G. (1999). The origin of war: Evolution of intergroup agonistic behavior in animals and men. Paper presented at the 19th Annual Conference of the Association for Politics and the Life Sciences, Atlanta, 2–5 September.

Van Vugt, M., & Ahuja, A. (2011). *Naturally selected: The evolutionary science of leadership*. New York, NY: Harper Collins.

Vanhanen, T. (1992). *On the evolutionary roots of politics*. New Delhi: Sterling Publishers.

Vanhanen, T. (2003). *Democratization: A comparative analysis of 170 countries*. London: Routledge.

Vanhanen, T. (2009). *The limits of democratization: Climate, intelligence, and resources*. Augusta, GA: Washington Summit Publishers.

Vanhanen, T. (2013). On the evolutionary roots of the social sciences. In S. A. Peterson & A. Somit (Eds.), *The world of biology and politics: Organization and research areas* (Vol. 11, pp. 45–70). Research in Biopolitics. Bingley, UK: Emerald Group Publishing Limited.

Vanhannen, T. (1984). *The emergence of democracy: A comparative study of 119 states, 1850–1979*. Helsinki: Finnish Society of Sciences and Letters.

Veblen, T. (2007). *The theory of the leisure class*. New York, NY: Oxford University Press.

Venables, N. C., Patrick, C. J., Hall, J. R., & Bernat, E. M. (2011). Clarifying relations between dispositional aggression and brain potential response: Overlapping and distinct contributions of impulsivity and stress reactivity. *Biological Psychology*, *86*, 279–288.

Verhulst, B., Eaves, L. J., & Hatemi, P. K. (2012). Correlation not causation: The relationship between personality traits and political ideologies. *American Journal of Political Science*, *56*, 34–51.

Verhulst, B., Hatemi, P. K., & Martin, N. G. (2010). The nature of the relationship between personality traits and political attitudes. *Personality and Individual Differences*, *49*, 306–316.

Vytal, K., & Hamann, S. (2010). Neuroimaging support for discrete neural correlates of basic emotions: A voxel-based meta-analysis. *Journal of Cognitive Neuroscience*, *22*(12), 2864–2885.

Wachtel, P. (1998). Overconsumption. In R. Keil, D. Bell, L. Fawcett, & P. Penz (Eds.), *Political ecology: Global and local* (pp. 259–271). London: Routledge.

Wahlke, J. C., & Lodge, M. G. (1972). Psychophysiological measures of political attitudes and behavior. *Midwest Journal of Political Science*, *16*(4), 505–537.

Wahlke, J. C. (1977). Research prospects for a biobehavioral political science. In *Conference of the Research Committee on Biology and Politics, International Political Science Association,* Bellagio, Italy, 12, 17.

Wahlke, J. C. (1979). Pre-behavioralism in political science. *American Political Science Review, 73*(1), 9–31.

Wahlke, J. C. (1991). Liberal learning and the political science major: A report to the profession. *Political Science & Politics, 24*(March), 48–60.

Wahlund, K., & Kristiansson, M. (2009). Aggression, psychopathy and brain imaging – Review and future recommendations. *International Journal of Law and Psychiatry, 32,* 266–271.

Wallace, B. (2010). *Getting Darwin wrong: Why evolutionary psychology won't work.* Exeter: Imprint Academic.

Wallerstein, I. (1974). *The modern world-system, Volume 1. Capitalist agriculture and the origins of the European world-economy in the sixteenth century.* New York, NY: Academic Press.

Wallerstein, I. (1979). *The capitalist world-economy.* Cambridge: Cambridge University Press.

Wallerstein, I. (1980). *The modern world-system, Volume 2. Mercantilism and the consolidation of the European world-economy, 1600–1750.* New York, NY: Academic Press.

Wallerstein, I. (1984). *The politics of the world-economy: The states, the movements, and the civilizations.* Cambridge: Cambridge University Press.

Wallerstein, I. (1989). *The modern world-system, Volume 3. Second era of great expansion of the capitalist world-economy.* San Diego, CA: Academic Press.

Wallerstein, I. (1991). *Unthinking social science: The limits of nineteenth century paradigms.* Cambridge: Polity Press.

Walsh, A. (1995). *Biosociology: An emerging paradigm.* Westport, CT: Praeger.

Warneken, F., & Tomasello, M. (2009). Cognition for culture. In P. Robbins & M. Aydede (Eds.), *Cambridge handbook of situated cognition* (pp. 467–479). Cambridge: Cambridge University Press.

Wasby, S. L., Peterson, S., Schubert, J., & Schubert, G. (1992). Supreme court's use of per curiam dispositions: The connection to oral argument. *The University of Illinois Law Review, 13,* 1.

Watson, D., Clark, L. A., & Tellegen, A. (1988). Development and validation of brief measures of positive and negative affect: The PANAS scales. *Journal of Personality and Social Psychology, 54*(6), 1063–1070.

Watts, M. (1994). Biopolitical contributions to comparative politics. In A. Somit & S. A. Peterson (Eds.), *Research in biopolitics* (Vol. 2). London: JAI Press.

Weaver, J. (2010). Amygdala at the centre of your social network. *Nature.* doi:10.1038/news.2010.699

Weber, C., Johnson, M., & Arceneaux, K. (2011). Genetics, personality, and group identity. *Social Science Quarterly, 92*(5), 1314–1337.

Weeden, J., Cohen, A. B., & Kenrick, D. T. (2008). Religious attendance as reproductive support. *Evolution and Human Behavior, 29*(5), 327–334.

Weinberger, J., & Westen, D. (2008). RATS, we should have used Clinton: Subliminal priming in political campaigns. *Political Psychology, 29*(5), 631–651.

Weinschenk, A. C. (2014). Personality traits and the sense of civic duty. *American Politics Research, 42*(1), 90–113.

Weisberg, H. F., & Nawara, S. P. (2010). How sophistication affected the 2000 presidential vote: Traditional sophistication measures versus conceptualization. *Political Behavior, 32*(4), 547–565.

Westen, D. (2007). *The political brain*. Jackson, TN: Public Affairs.

Westen, D., Blagov, P. S., Harenski, K., Kilts, C., & Hamann, S. (2006). Neural bases of motivated reasoning: An FMRI study of emotional constraints on partisan political judgment in the 2004 U.S. Presidential election. *Journal of Cognitive Neuroscience, 18*(11), 1947–1958.

White, E. (1972). Genetic diversity and political life: Toward a populational-interaction paradigm. *The Journal of Politics, 34*(4), 1203–1242.

White, E. (1992). *The end of the empty organism: Neurobiology and the sciences of human action*. Westport, CT: Praeger.

White, E. (1993). Genes, brains, and politics: Human evolution, behavior, and *i*ntelligence. Westport, CT: Praeger.

White, E. (1997). The balkanized global village: A psychogenetic perspective. In A. Somit & S. A. Peterson (Eds.), *Research in biopolitics* (Vol. 5, pp. 71–102). London: JAI Press.

Whitehead, A. N. (1929). *The function of reason*. Princeton, NJ: Princeton University Press.

Wiarda, H. (Ed.). (1985). *New directions in comparative politics*. Boulder, CO: Westview Press.

Wiegele, T. C. (1973). Decision making in an international crisis: Some biological factors. *International Studies Quarterly, 17*(3), 295–335.

Wiegele, T. C. (1976). Health and stress during international crisis: Neglected input variable in the foreign policy decision-making process. *Journal of Political Science, 3*, 139–144.

Wiegele, T. C. (1977a). Models of stress and disturbances in elite political behaviors: Psychological variables and political decision making. In R. S. Robins (Ed.), *Psychopathology and political leadership* (Vol. 16, pp. 79–111). New Orleans, LA: Tulane Studies in Political Science.

Wiegele, T. C. (1977b). Physiologically-based content analysis: An application in political communication. In B. Rubin (Ed.), *Communication yearbook II*. New Brunswick, NJ: Transaction Books.

Wiegele, T. C. (1978). The psychophysiology of elite stress in five international crises: A preliminary test of a voice measurement technique. *International Studies Quarterly, 22*, 467–511.

Wiegele, T. C. (1979a). *Biopolitics: Search for a more human political science*. Boulder, CO: Westview Press.

Wiegele, T. C. (1979b). Signal leakage and the remote psychological assessment of foreign policy elites. In L. Falkowski (Ed.), *Psychological models in international politics*. Boulder, CO: Westview Press.

Wiegele, T. C. (1982). The case for a biological perspective in the study of international relations. In G. Hopple (Ed.), *Biopolitics, political psychology and international relations* (pp. 151–173). London: Francis Pinter.

Wiegele, T. C., Plowman, S. A., & Caty, R. (1975). Cardiorespiratory health and dimensions of subject attitudes toward international affairs. *Experimental Study of Politics, 4*(1), 36–54.

Wilson, E. O., & Wilson, D. S. (2007). Rethinking the theoretical foundation of sociobiology. *Quarterly Review of Biology, 82*, 327–348.

Wilson, E. O. (1975). *Sociobiology: The new synthesis*. Cambridge, MA: Belknap.

Wilson, E. O. (1977). Biology and the social sciences. *Daedalus, 106*(4), 127–140.

Wilson, E. O. (Ed.). (1988). *Biodiversity*. Washington, DC: National Academy of Sciences/ Smithsonian Institution.

Wilson, J. Q. (1993). *The moral sense*. New York, NY: Free Press.

Wilson, E. O. (1998). *Consilience: The unity of knowledge*. New York, NY: Vintage.

Wilson, M. (2010). The re-tooled mind: How culture re-engineers cognition. *Social Cognitive and Affective Neuroscience*, *5*(2–3), 180–187.

Wilson, E. O. (2012). *The social conquest of earth*. New York, NY: Liveright Publishing Corporation.

Witter, R., Stinson, K., Sackett, H., Putter, S., Kinney, G., Teitelbaum, D., & Newman, L. (2008). *Potential exposure-related human health effects of oil and gas development: A white paper*. NRDC. Retrieved from http://docs.nrdc.org/health/files/hea 080091701702a.pdf. Accessed on September 15.

Wolak, J., & Marcus, G. E. (2007). Personality and emotional response: Strategic and tactical responses to changing political circumstances. *The Annals of the American Academy of Political and Social Science*, *614*(1), 172–195.

Wolin, S. (1960). *Politics and vision*. Boston, MA: Little Brown.

Woodhouse, E. (1972). Re-visioning the future of the third world: An ecological perspective on development. *World Politics*, *25*(1), 1–33.

Wright, Q. (1955). *The study of international relations*. New York, NY: Appleton-Century-Crofts.

Wright, Q. (1965). *A study of war* (2nd ed.). Chicago, IL: University of Chicago Press.

Wright, R. (1994). *The moral animal: Evolutionary psychology and everyday life*. New York, NY: Pantheon Books.

Yergin, D., & Stanislaw, J. (2002). *The commanding heights: The battle for the world economy*. New York, NY: Touchstone.

Yergin, D. (2009). *The prize: The epic quest for oil, money and power*. New York, NY: Free Press.

Yergin, D. (2011). *The quest: Energy, security, and the remaking of the modern world*. New York, NY: Penguin Press.

Zahavi, A., & Zahavi, A. (1997). *The handicap principle: A missing piece of Darwin's puzzle*. New York, NY: Oxford University Press.

Zak, P. J., & Denzau, A. T. (2001). Evolutionary approaches in the behavioral sciences: Toward a better understanding of human nature. In A. Somit (Ed.), *Evolutionary approaches in the behavioral sciences: Toward a better understanding of human nature* (pp. 31–65). Greenwich, CT: JAI.

Zamboni, G., Gozzi, M., Krueger, F., Duhamel, J. R., Sirigu, A., & Grafman, J. (2009). Individualism, conservatism, and radicalism as criteria for processing political beliefs: A parametric FMRI study. *Social Neuroscience*, *4*(5), 367–383.

Zaretsky, M., Mendelsohn, A., Mintz, M., & Hendler, T. (2010). In the eye of the beholder: Internally driven uncertainty of danger recruits the amygdala and dorsomedial prefrontal cortex. *Journal of Cognitive Neuroscience*, *22*(10), 2263–2275.

Zhou, W., Sornette, D., Hill, R. A., & Dunbar, R. I. (2005). Discrete hierarchical organization of social group sizes. *Proceedings of the Royal Society B: Biological Sciences*, *272*(1561), 439–444.

Zink, C. F., Pagnoni, G., Chappelow, J., Martin-Skurski, M., & Berns, G. S. (2006). Human striatal activation reflects degree of stimulus saliency. *NeuroImage*, *29*, 977–983.

Zuckerman, M. (1994). *Behavioral expressions and biosocial bases of sensation seeking*. Cambridge: Cambridge University Press.

ABOUT THE AUTHORS

Robert H. Blank (Ph.D., University of Maryland) is a Research Scholar at the New College in Florida, USA, and an Adjunct Professor at the University of Canterbury in Christchurch, New Zealand. He is also a frequent Guest Professor at Aarhus University in Denmark. Previous academic positions include Chair of Public Policy at Brunel University in West London, and Professor and Associate Director of the Program for Biosocial Research at Northern Illinois University. He has also held residential fellowships at Vanderbilt University, Stanford University, Indiana University, and the Medical University of South Carolina, and Fulbright Lectureships in Taiwan (1976–1977 and 2007) and New Zealand (1984). He has taught, lectured, and written widely in the areas of comparative health policy, medical technology assessment, genetic and reproductive policy, and biomedical ethics/policy. Among the many books he has written or co-authored are *Rationing Medicine* (Columbia University Press), *Regulating Reproduction* (Columbia University Press), *Biomedical Policy* (Nelson-Hall), *Brain Policy* (Georgetown University Press), *The Price of Life* (Columbia University Press), *Biology and Political Science* (Routledge), *Comparative Health Policy* (Palgrave), *End of Life Decision Making: A Comparative Study* (MIT Press), *Condition Critical* (Fulcrum Press), and *Intervention in the Brain: Politics, Policy and Ethics* (MIT Press).

Odelia Funke (Ph.D., University of Virginia) is a Senior Advisor in the Office of Environmental Information, US Environmental Protection Agency (EPA). At EPA, she has worked in a variety of positions as a senior analyst, senior advisor, and manager, covering a wide array of environmental and information issues. Before joining EPA, she was an Assistant Professor, teaching graduate and undergraduate courses in political theory and government. She has been an Adjunct Professor and Lecturer at several local universities during her EPA tenure, teaching courses in environmental politics and ethics, women and politics, government, and privacy. She has been a Guest Lecturer, and written conference papers, peer review articles, and book chapters on biopolicy, environmental policy, environmental security, health policy and privacy, bioterrorism, and political theory. Most recently, she spent a year as an Executive in Residence at American University's Center for Environmental Policy, School of Public Affairs, doing research on governance issues related to hydraulic fracturing of natural gas and teaching.

Samuel M. Hines, Jr. (Ph.D., Duke University) is the Provost and Dean of the College and Professor of Political Science at The Citadel: The Military College of South Carolina. Previously, he was Vice President for Departmental Affairs and Associate Provost as well as Dean of the School of Humanities and Social Sciences at the College of Charleston. He has authored numerous articles, books, and book chapters, and was the editor of *South Atlantic Urban Studies: An Urban Affairs Annual*. He served two terms as Chairman of the Executive Council of the Association for Politics and the Life Sciences and continues to serve as a member of the Executive Council. He has received research and project grants totaling over one million dollars from the National Institute on Aging, the Department of Defense/Office of Naval Reserve, the National Endowment for the Humanities, the U.S. Department of Education, and other foundations. His principal research interests are in the fields of biopolitics, leadership, entrepreneurship studies, political theory, public administration, and international politics. In the field of biopolitics, he has written seminal articles on biopolitical theory, political evolution, international political development, and age and political behavior. His most recent book (co-authored with Robert Blank) is *Biology & Political Science* (Routledge).

Joseph Losco (Ph.D., Temple University) is Professor of political science at Ball State University where he teaches political theory and American government. He is Co-Director of the Bowen Center for Public Affairs where he shares responsibility for conducting the annual Hoosier Survey on Public Policy and conducting research into electoral systems. He publishes in the fields of political theory, public policy, and biopolitics. His publications in biopolitics include *Human Nature and Politics*, co-edited with Albert Somit (JAI Press); "Evolutionary Theory and Political Thought" (in S. Peterson and A. Somit (Eds.), *Evolutionary Approaches in the Behavioral Sciences*); "Paternal-Fetal Conflict: An Examination of Paternal Responsibilities to the Fetus" (*Politics and the Life Sciences*); and "From Outrage to Orthodoxy? Sociobiology and Political Science at 35" (*Politics and the Life Sciences*). He is co-author (with Ralph Baker) of *AMGOV*, an annually updated text in American Government and *Political Theory: Classic Writings and Contemporary Views* (co-authored with Leonard Williams) published by Oxford Press. He has also published in the area of higher education policy including *Higher Education in Transition: Challenges of the New Millennium*, co-edited with Brian Fife (Bergin & Garvey, 2000).

Patrick Stewart (Ph.D., Northern Illinois University) is an Associate Professor at the University of Arkansas, Fayetteville. In addition to his book *Debatable Humor: Laughing Matters on the 2008 Presidential Primary Campaign* (2012), he has published research concerning nonverbal communication by politicians in the journals *PLoS-ONE, Political Psychology, Motivation and Emotion, International Journal of Humor Research, PS – Political Science & Politics,* and *Politics and the Life Sciences*. His work on nonverbal communication by political leaders has been reported on by such popular outlets as *The Washington Post, U.S. News and World Report, Wired, Psychology Today, The National Review, The Huffington Post*, amongst others. He is a certified Facial Action Coding System (FACS) coder whose current research concentrates on the emotional response of followers to leaders (and potential leaders). Specifically, he is researching audience laughter and applause in response to comments by candidates during presidential debates, how individuals respond to different types of facial displays by presidential candidates, and how the emotion of disgust influences political attitudes and opinions.